Lecture Notes
in Business Information Processing   **522**

LNBIP reports state-of-the-art results in areas related to business information systems and industrial application software development – timely, at a high level, and in both printed and electronic form.

The type of material published includes

- Proceedings (published in time for the respective event)
- Postproceedings (consisting of thoroughly revised and/or extended final papers)
- Other edited monographs (such as, for example, project reports or invited volumes)
- Tutorials (coherently integrated collections of lectures given at advanced courses, seminars, schools, etc.)
- Award-winning or exceptional theses

LNBIP is abstracted/indexed in DBLP, EI and Scopus. LNBIP volumes are also submitted for the inclusion in ISI Proceedings.

Simone Agostinelli

Generating Executable Robotic Process Automation Scripts from Unsegmented User Interface Logs

 Springer

Simone Agostinelli
Sapienza University of Rome
Rome, Italy

ISSN 1865-1348   ISSN 1865-1356 (electronic)
Lecture Notes in Business Information Processing
ISBN 978-3-031-61367-8   ISBN 978-3-031-61368-5 (eBook)
https://doi.org/10.1007/978-3-031-61368-5

This Springer imprint is published by the registered company Springer Nature Switzerland AG
The registered company address is: Gewerbestrasse 11, 6330 Cham, Switzerland

There's nothing we could do,
because this world is just that cruel.
† Berthold Huber, SNK

Preface

Business Processes (BPs) are nowadays an integral part of mid-size and large organizations that aim to ensure consistent business outcomes and take advantage of improvement opportunities to remain competitive [88]. Examples of traditional BPs include insurance claim processing, order handling, and sales management. Business Process Management (BPM) is the discipline that oversees how BPs are performed in an organization, providing concepts and tools to support the design, administration, enactment, and analysis of BPs.

Since the late 1990s, a new generation of information systems, called Business Process Management Systems (BPMSs), have become increasingly popular to automate running BPs involving people, applications, and information sources based on BP specifications (i.e., process models) pre-defined at design-time [106]. BPMSs seek to improve the efficiency of BPs by streamlining their execution through the orchestrated distribution of work items to process participants and software services, thus reducing the time required to run the everyday operations [36].

However, automating a BP specification using a BPMS requires a not negligible development effort that involves dedicated technical resources, which are in charge of specifying the execution properties (many of them are vendor-specific) of each BP element and the connectors to the Application Programming Interfaces (API) of the various applications that realize the behaviour of the BP. In addition, due to an acceleration of the digital transformation process enacted by many organizations, the number of BPs to manage and execute in organizations through a BPMS is constantly growing over the years [36, 48, 103]. For this reason, BPMSs are turning out to be too inflexible for fast and lightweight automation projects, where the investment to implement and maintain the automated BPs may exceed the manual costs of operation [46].

To mitigate this issue, Robotic Process Automation (RPA) is a maturing automation technology in the field of BPM that creates software (SW) robots to partially or fully automate rule-based and repetitive tasks (or simply *routines*) performed by human users in their applications' user interfaces (UIs) [103]. RPA is thought to provide the shortest route to business process (BP) automation by accessing only the

UI layer of IT systems rather than going deeply into the application code or databases sitting behind them [84].

In recent years, much progress has been made both in terms of research and technical development on RPA, resulting in many industry-specific deployments for industrial-oriented services [57, 12, 14, 55, 91, 92, 58]. Moreover, the market of RPA solutions has developed rapidly and today includes more than 50 vendors developing tools that provide SW robots with advanced functionalities for automating office tasks of different complexity [13]. Nonetheless, when considering state-of-the-art RPA technology, it becomes apparent that the current generation of RPA tools is driven by predefined rules and manual configurations made by expert users rather than automated techniques [9, 10, 25].

To be more specific, the traditional workflow to conduct an RPA project can be summarized as follows [51]:

1. Determine which routines are good candidates to be automated.
2. Record the mouse/key events that happen on the UI of the SW applications involved in a routine execution, i.e., the *UI logs*.
3. Model the selected routines in the form of flowchart diagrams, which involve the specification of the actions, routing constructs (e.g., parallel and alternative branches), data flow, etc. that define the behaviour of a SW robot.
4. Develop each modeled routine by generating the SW code required to concretely enact the associated SW robot on a target computer system.
5. Deploy the SW robots in their environment to perform their actions.
6. Monitor the performance of SW robots to detect bottlenecks and exceptions.
7. Maintain the routines, which take into account the SW robot's performance and error cases to eventually enhance their behaviour.

The majority of the previous steps, particularly the ones involved in the early stages of the RPA life-cycle, require the support of skilled human experts, who need to: *(i)* understand the anatomy of the candidate routines to automate by means of interviews, walk-troughs, and detailed observation of workers conducting their daily work (cf. step 1); and *(ii)* define manually the flowchart diagrams representing the structure of such routines (cf. step 3), which will drive the development of the SW code, often in form of executable scripts (also called *RPA scripts*), allowing the concrete enactment of SW robots at run-time (cf. step 4). While this approach is effective in executing simple rules-based logic in situations where there is no room for interpretation, it becomes time-consuming and error-prone in the presence of routines that are less predictable or require some level of human judgment [74, 10]. Indeed, the designer should have a global vision of all possible variants of the routines to define the appropriate behaviours of the SW robot, which becomes complicated when the number of variants increases. The issue is that in case where the flowchart diagram does not contain a suitable response for a specific situation, e.g., because of a shallow modeling activity, then the associated RPA scripts would not properly reflect the behaviour of the potential routine variant, forcing SW robots to escalate to a human supervisor at run-time, in contrast with the RPA philosophy.

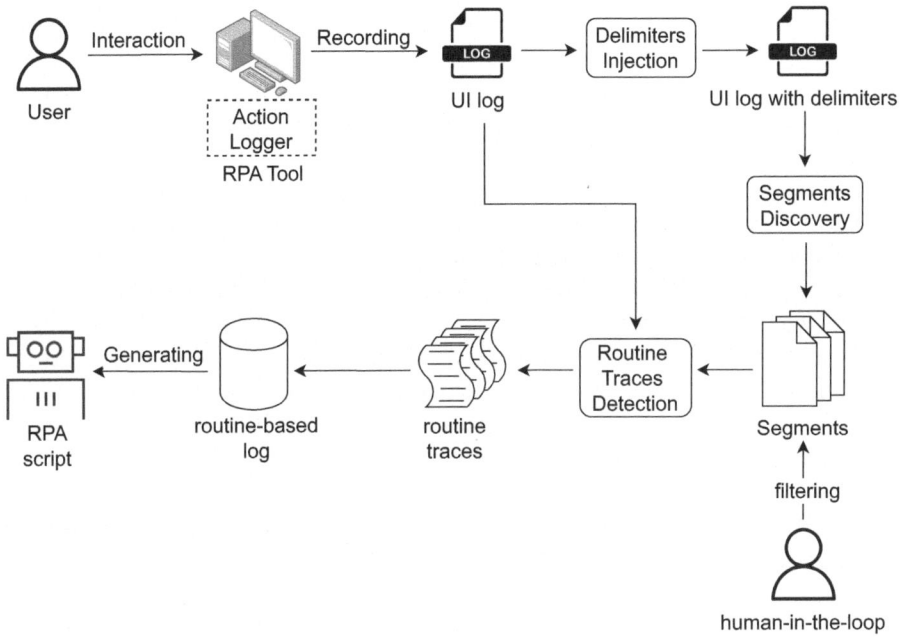

Fig. 1 Overview of the adopted approach

To address the above issues, and mitigate the involvement of skilled human experts in steps 1, 3 and 4, the thesis underlying this book is targeted to: *(i)* automatically understand which user actions contribute to which routines inside a UI log (this issue is known as *segmentation*) and *(ii)* automatically generate executable RPA scripts directly from the UI logs that record the user interactions with the SW applications involved in a routine execution, thus skipping completely the (manual) modeling activity of the flowchart diagrams.

Although RPA is generally considered an easy-to-implement technology, in-depth knowledge is necessary to create reliable and scalable SW robots, particularly when the intervention of human experts is required to properly progress the execution of a routine. As a result, between 30% and 50% of initial RPA implementations are estimated to fail [86, 58]. Consequently, an approach that simplifies the realization of an RPA project towards the automated identification of user actions belonging to a specific routine inside a UI log, with the subsequent generation and enactment of the associated SW robot, can be considered a relevant artefact to investigate.

To achieve these goals, as shown in Fig. 1, starting from an unsegmented UI log previously recorded with an RPA tool, the first stage of this research is to inject into the UI log the *end-delimiters* of the routines under examination. An end-delimiter is a dummy action added to the UI log immediately after the user action that is known to complete a routine execution. The knowledge of such end-delimiters is crucial to make the approach work, as discussed later in the book.

The second step of the approach is to automatically discover the most frequently observed segments of the routines as recorded in the UI log with the end-delimiters. *Routine segments* describe the different behaviours of the routine(s) under analysis, in terms of repeated patterns of performed user actions. An unsegmented UI log consists of sequential data of user actions performed on the UI of a computer system during many routines' executions. To automatically identify the routine segments from the UI log, we rely on a *frequent-pattern identification technique* [27], which we customize on an ad-hoc basis. In this phase, the risk exists that some wrong routine segments are discovered, i.e., not allowed from the real-world routines that are known to be valid at the outset.

To address this issue, a *human-in-the-loop interaction* that enables human experts to visualize the *declarative constraints* inferred by the discovered routine segments is required. Such constraints describe the temporally extended relations between user actions that must be satisfied throughout a routine segment (e.g., an action a_1 must be eventually followed by an action a_2). In a nutshell, they collectively determine the observed behaviours of the routine segments from the UI log. This knowledge allows human experts to identify and remove those constraints that should not be compliant with any real-world routine behaviour, thus filtering out the not valid (i.e., wrongly discovered) routine segments.

Finally, starting from any of the remaining (valid) routine segments, a customized version of a *trace alignment* technique in Process Mining [2, 30] is employed to automatically detect and extract the *routine traces* by the original UI log. A routine trace represents an execution instance of a routine within a UI log. By identifying the routine traces, it is also possible to filter out those actions in the UI log that are not part of the routine under observation and hence are redundant or represent noise. Such traces are finally stored in a dedicated *routine-based log*, which captures exactly all the user actions that happened during many different executions of the routine, thus achieving the segmentation task (cf. step 1). Therefore, the final outcome of our segmentation approach will be a collection of as many routine-based logs as the number of valid routine segments.

The customized version of the trace alignment technique can also be used as a supervised segmentation technique [11]. The *supervised* assumption, which consists of knowing a priori the structure of routines, may ease the segmentation task. Still, as a side effect, it may strongly constrain the discovery of routine traces only to the "paths" allowed by the routines' structure, thus neglecting that some valid yet infrequent routine variants may exist in the UI log. For this reason, the novelty of the proposed approach to the segmentation of UI logs [3] is to semi-automatically discover such structures in the form of routine segments and then use them as input for the supervised segmentation technique [11].

Most state-of-the-art segmentation approaches can properly extract routine segments from unsegmented UI logs when the routine executions are not interleaved from each other. Only a few works are able to partially untangle unsegmented UI logs of many interleaved routine executions, but with the assumption that any routine provides its own, separate universe of user actions. This is a relevant limitation since it is quite common that real-world routines may share the same user actions (e.g., copy/paste data across cells of a spreadsheet) to achieve their objectives. In this book, we propose an interactive approach to the segmentation of UI logs [3] that aims to mitigate the issues mentioned above, showing its ability to outperform existing literature approaches in terms of supported segmentation variants and to which extent the human-in-the-loop interaction can filter out the wrongly discovered segments. In particular, we evaluated the *robustness* of the proposed approach to (re)discover the valid routine segments from synthetic UI logs of increasing complexity. Then, we also investigated the degree of *effectiveness*, *robustness*, and *usability* of the tool implementing the human-in-the-loop interaction step.

It is worth noticing that commercial RPA tools can eventually employ routine-based logs to synthesize executable scripts in the form of SW robots that will emulate the routine behaviour on the UI without the manual modeling of the routines (cf. steps 3 and 4). In this direction, a cross-platform software tool called SmartRPA[1][4, 5] was developed, which is able to generate executable RPA scripts, necessary to enact the SW robot that implements the selected routine variant directly from a segmented UI log (i.e., a routine-based log). A *routine variant* is a specific execution of a routine that differs from the other executions (i.e., instances) of the same routine in at least one user action. Differently from the literature approaches to automated RPA script generation from UI logs, which enable the automation of straightforward routines that have essentially no variance and do not require any human intervention, the SW robots generated by SmartRPA are obtained to handle the intermediate user inputs that are required during the routine execution, thus enabling to emulate the most suitable routine variant for any specific combination of user inputs as observed in the UI log. This makes the synthesis of SW robots performed by SmartRPA *reactive* to any user decision found during a routine execution. "Reactivity" highlights the fact that the behaviour of SW robots is determined immediately before their enactment, as it is driven by the specific user inputs required to execute the routine. This also means that reactivity enables the potential run-time generation of as many SW robots as the routine variants to be emulated.

SmartRPA has been validated on four non-functional requirements to measure the quality of the underlying approach. Specifically, we first performed many synthetic experiments employing UI logs of increasing complexity to assess the *robustness* and *feasibility* of SmartRPA to the identification of routine variants and variation points for the reactive synthesis of SW robots. A *variation point* is a point in the routine execution where a user choice needs to be made between multiple possible routine variants. Then, we performed a controlled experiment involving real users exploiting our RPA use case to investigate the *effectiveness* of the SmartRPA approach [6] when compared to a traditional model-based approach for the generation of SW robots. Finally, we quantify the *usability* of the UI provided by the tool implementing the SmartRPA approach.

In summary, the thesis underlying this book tries to mitigate the involvement of skilled human experts in steps 1, 3, and 4, throughout the development of: *(i)* an interactive approach to the automated segmentation of UI logs, and *(ii)* the SmartRPA

---

[1] SmartRPA is available at: `https://github.com/bpm-diag/smartRPA`

approach to the automated identification of the variation points of a routine, to enable the selection of the most suitable routine variants to be implemented with a SW robot directly from a routine-based log.

Book Structure

The book is organized into 9 chapters and organized into 4 parts, as follows:

- Part I paves the way for both the automated segmentation of UI logs and the automated generation of SW robots, laying the foundations for understanding both topics and targeting a broader audience in the context of BPM.

  - Chapter 1 reports an introduction that makes immediately clear for the reader the research problem addressed, its significance in the RPA field, and the proposed contribution to solving the problems, driven by specific research challenges. This serves as the basis for positioning the performed work and summarizing the author's research activities.
  - Chapter 2 analyzes the background notions about BPM in general, RPA and Process Mining. Specifically, it outlines the preliminaries on routines, SW robots, UI logs, Petri nets, Trace Alignment and DECLARE. Furthermore, it presents a real-life RPA use case to illustrate the relevance of the research challenges being investigated.

- Part II focuses on the issue of automated segmentation of UI logs in RPA. It presents an approach to tackle such an issue and its implementation.

  - Chapter 3 focuses on the issue of segmentation of UI logs, identifying all its potential variants and presenting an up-to-date overview that discusses to what extent existing literature approaches support such variants.
  - Chapter 4 presents the employed technique for discovering routine segments directly from unsegmented UI logs, that is, a frequent-pattern identification technique (properly customized for our purposes) to automatically derive the routine segments as recorded into a UI log. We evaluated the robustness of this technique in the presence of synthetic UI logs of a growing size that provide an increasing amount of routine variants to measure to what extent our approach is able to (re)discover the valid routine segments from such UI logs.
  - Chapter 5 implements the human-in-the-loop interaction step to filter out those segments not allowed (i.e., wrongly discovered from the UI log) by any real-world routine under analysis. We have also evaluated the implemented technique by measuring its degree of effectiveness, robustness, and usability.
  - Chapter 6 presents the routine traces detection component which exploits trace alignment in Process Mining to extract from a UI log all those user actions belonging to a specific (valid) routine segment and cluster them into well-bounded routine traces, thus achieving the segmentation task. It is worth noticing that this component can also be employed as a stand-alone supervised

segmentation technique, under the assumption to know a priori the structure (i.e., the flowchart diagrams) of the routines to identify in the UI log, thus neglecting the semi-automated discovery of the routine segments.

- Part III focuses on the automated generation of SW robots by means of executable scripts through the implementation of the SmartRPA approach.

  - Chapter 7 leverages a design science research method [52] to develop an approach, called SmartRPA, which is able to interpret the UI logs keeping track of many routine executions and automatically synthesize SW robots that emulate the most suitable routine variant for any specific intermediate user input that is required during the routine execution. Specifically, it is focused on *(i)* discussing the relevant state-of-the-art approaches that attempt to mitigate the research challenges, *(ii)* deriving a set of technical requirements to realize our SmartRPA approach, *(iii)* proposing the SmartRPA approach and describing its stages to address the technical requirements, and finally *(iv)* presenting the details of an algorithm to automatically identify routine variants and variation points from UI logs, necessary for the reactive synthesis of SW robots.
  - Chapter 8 shows the technical steps enacted to develop the SmartRPA approach as a real implemented tool and presents the results of a multi-step evaluation performed on SmartRPA to investigate the extent to which the approach satisfies four relevant non-functional requirements, namely robustness, feasibility, effectiveness and usability employing both synthetic and real-world datasets.

- Part IV concludes the book.

  - Chapter 9 concludes the book by discussing limitations and future developments. Moreover, it summarizes the results, impacts and benefits addressed by this book.

The present book is strongly based on the PhD dissertation of Simone Agostinelli, which was submitted to the Sapienza University of Rome in January 2022 and defended in May 2022. Fig. 2 provides a roadmap of the book with possible paths for the reader to follow. The numbers in the nodes of the graph represent the chapters. The blue path is recommended for readers interested in the automated segmentation of UI logs, while the red path is suggested for those interested in the automated generation of SW robots through executable RPA scripts. However, all chapters can be sequentially pursued by following the green path since they are self-contained.

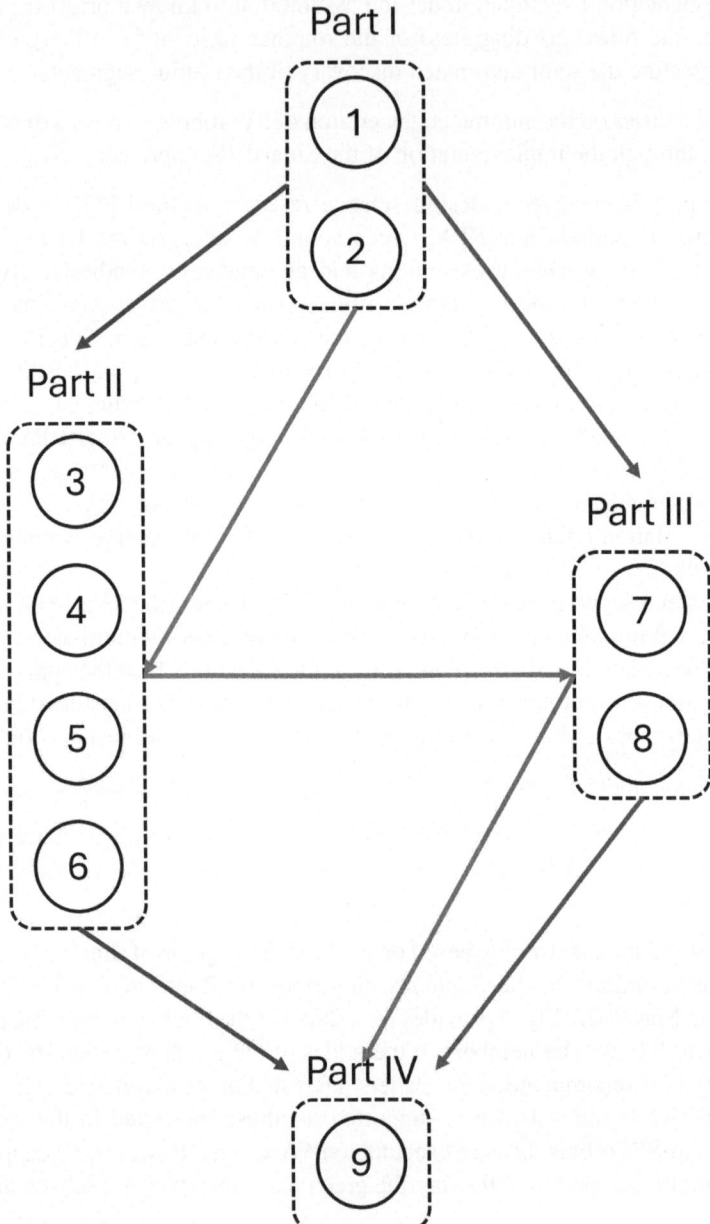

Fig. 2 Road map of the book

Acknowledgements

The PhD experience has made me a better person both academically and personally. Constantly challenging myself on my choices has been the best lesson this journey could leave me with. If I have succeeded in this path, it is not only thanks to my perseverance but also thanks to my supervisor's guidance. Andrea Marrella was not just my PhD supervisor; he has been something no one else has ever been to me. A figure whom I esteem and respect, always ready to show me the way if I was going down the wrong path or to support me if I was about to give up.

I also thank Massimo Mecella, Francesco Leotta, and all the members of my university research group with whom I have worked and shared ideas. Beyond my university, I also extend my thanks to Tom Hohenadl from KU Eichstätt-Ingolstadt (Germany) and Antonio Martínez-Rojas from Universidad de Sevilla (Spain) for their collaboration on SmartRPA and for the new research directions we are exploring on Robotic Process Automation.

I would also like to thank the reviewers of my PhD dissertation, Hajo A. Reijers from Utrecht University (Netherlands) and Ernest Teniente from Universitat Politècnica de Catalunya (Spain), for the time and effort dedicated to providing careful reviews. Their insightful, constructive, and meticulous comments have strongly contributed to improving the quality of my work.

Special mention goes to my girlfriend, Silvia, for her unconditional and loving support that sustains me in every situation.

Last but not least, I am grateful to my family for having made all of this possible, and to my old, lost friends, with whom I have shared a great part of my life and whom I thank for shaping the person I have become today.

Contents

Acronyms

AI	Artificial Intelligence
ABPMS	AI-augmented Business Process Management Systems
API	Application Programming Interfaces
BNF	Backus Normal Form
BP	Business Process
BPM	Business Process Management
BPMN	Business Process Model and Notation
BPMS	Business Process Management Systems
CRM	Customer Relationship Management
CSP	Communicating Sequential Process
CSV	Comma-Separated Values
CTT	Concur Task Trees
DFG	Directly-Follows Graph
EPC	Event-driven Process Chain
ERP	Enterprise Resource Planning
GOMS	Goals Operators Methods Selection
GUI	Graphical User Interface
HTML	HyperText Markup Language
LTL	Linear Temporal Logic
LLM	Large Language Models
MDL	Minimum Description Length
JSD	Jackson System Development
OCEL	Object-Centric Event Logs
PMS	Process Management System
RPA	Robotic Process Automation
RPM	Robotic Process Mining
STN	State Transition Network
SW	Software
UI	User Interface
URL	Uniform Resource Locator
XAML	Extensible Application Markup Language

XES   eXtensible Event Stream
XML   Extensible Markup Language
YAWL   Yet Another Workflow Language

Part I
Prerequisites

Chapter 1
Introduction

Business Processes (BPs) are nowadays an integral part of mid-size and large organizations that aim to ensure consistent business outcomes and take advantage of improvement opportunities to remain competitive [88]. Traditional BPs include insurance claim processing, order handling, and sales management. Business Process Management (BPM) is the discipline that oversees how BPs are performed in an organization, providing concepts and tools to support the design, administration, enactment, and analysis of BPs.

Since the late 1990s, a new generation of information systems called Business Process Management Systems (BPMSs) has become increasingly popular to automate running BPs involving people, applications, and information sources based on BP specifications (i.e., process models) pre-defined at design time [106]. BPMSs seek to improve the efficiency of BPs by streamlining their execution through the orchestrated distribution of work items to process participants and software services, thus reducing the time required to run the everyday operations [36].

However, automating a BP specification using a BPMS requires a not negligible development effort that usually involves dedicated technical resources, which are in charge to specify the execution properties (many of them are vendor-specific) of each BP element and the connectors to the Application Programming Interfaces (API) of the various applications that realize the behaviour of the BP. The fact is that the number of BPs to manage and execute in organizations through a BPMS is constantly growing over the years [36, 48, 103], and BPMSs are turning out to be too inflexible for fast and lightweight automation projects, where the investment to implement the automated BPs may exceed the manual costs of operation [46].

To mitigate this issue, Robotic Process Automation (RPA) is a maturing automation technology in the BPM domain [54] that creates software (SW) robots to partially or fully automate rule-based and repetitive tasks (or simply *routines*) performed by human users in their applications' user interfaces (UIs) [103]. While conducting a BPM project is often considered too expensive because its "top-down" approach that forces to develop the PMS from scratch (and system integration is costly), RPA promises to rely on an approach where, instead of redesigning existing information systems (that remain unchanged), humans are replaced by SW robots in

S. Agostinelli: *Generating Executable Robotic Process Automation Scripts from Unsegmented User Interface Logs*, LNBIP 522, pp. 3–9, 2024.
https://doi.org/10.1007/978-3-031-61368-5_1

the execution of those BPs involving routine work. This allows knowledge workers to have more time for value-added tasks.

In the research literature, many case studies have shown that RPA technology can concretely lead to improvements in efficiency for BPs involving routine work in large companies, such as O2 and Vodafone [57, 12, 41]. Indeed, in recent years, much progress has been made both in terms of research and technical development on RPA, resulting in many industry-specific deployments for industrial-oriented services [57, 12, 14, 55, 91, 92, 58]. Moreover, the market of RPA solutions has developed rapidly. Today includes more than 50 vendors developing tools that provide SW robots with advanced functionalities for automating office tasks of different complexity [13].

However, despite this growing attention around RPA, to achieve more widespread adoption in the BPM domain, RPA needs to become "smarter" [103], i.e., RPA tools can adapt and learn how to handle non-standard cases by observing human problem resolving unexpected system behaviour (e.g., in case of system errors, changing forms, etc.). In fact, when considering the state-of-the-art technology, it becomes apparent that the current generation of RPA tools is driven by predefined rules and manual configurations made by expert users rather than by automated intelligent techniques [9, 10, 25]. Consequently, more complex and less defined BPs could be fully supported by the RPA technology. To be more specific, the traditional workflow to conduct an RPA project can be summarized as follows [51]:

1. Determine which process steps (also called *routines*) are good candidates to be automated.
2. Model the selected routines in the form of *flowchart diagrams* (i.e., the *interaction models*), which involve the specification of the actions, routing constructs (e.g., parallel and alternative branches), data flow, etc. that define the behaviour of a SW robot.
3. Record the mouse/key events that happen on the UI of the user's computer system. This information is associated with a routine's actions, enabling it to emulate the recorded human activities through a SW robot.
4. Develop each modeled routine by generating the software code required to concretely enact the associated SW robot on a target computer system.
5. Deploy the SW robots in their environment to perform their actions.
6. Monitor the performance of SW robots to detect bottlenecks and exceptions.
7. Maintenance of the routines, which takes into account each SW robot's performance and error cases. The outcomes of this phase enable a new analysis and design cycle to enhance the SW robots.

The majority of the previous steps, particularly the ones involved in the early stages of the RPA life-cycle, require the support of skilled human experts, which need to: *(i)* understand the anatomy of the candidate routines to automate through interviews, walk-troughs, and detailed observation of workers conducting their daily work (cf. step 1); and *(ii)* define manually the flowchart diagrams representing the structure of such routines (cf. step 3), which will drive the development of the SW code, often in the form of executable scripts (also called RPA scripts), allowing the concrete enactment of SW robots at run-time (cf. step 4).

Towards this direction, two research challenges necessary to inject intelligence into the current RPA technology towards better support to BPM can be derived, as discussed in [9, 10, 63, 8]:

1. **C1 - Automated Segmentation of UI Logs**.
   Description: UI logs recorded by RPA tools are characterized by long sequences of user actions that reflect many routine executions. A UI log can record information about several routines whose actions are mixed in some order that reflects the particular order of their execution by the user [22]. In addition, the same routine can be spread across multiple logs, interleaved with other actions that are not part of the routine under analysis (and potentially shared by many routines), making the automated identification of routines far from being trivial.
   Objective: Automatically identify and understand which user actions contribute to a particular routine inside a UI log (that keeps track of the user actions taking place during a run of the system) and cluster them into well-bounded routine traces (i.e., complete execution instances of a routine). This issue is known as "*segmentation*" (cf. step 1).

2. **C2 - Automated Generation of SW Robots**.
   Description: In RPA tools, there is a lacking of testing environments. As a consequence, SW robots are developed through a *trial-and-error* approach consisting of three steps that are repeated until success [62]: *(i)* First, a human designer produces a flowchart diagram (or an executable RPA script) that includes the actions to be performed by the SW robot on a target computer system at run-time; *(ii)* Second, SW robots are typically deployed in production environments, where they interact with information systems, with a high risk of errors due to inaccurate modeling of flowcharts; *(iii)* Third, if SW robots are not able to reproduce the behaviour of the users for a specific routine, then the designer adjusts the flowchart diagrams to fix the identified gap. While this approach is proven effective in executing simple rules-based logic in situations where there is no room for interpretation, it becomes time-consuming and error-prone in the presence of routines that are less predictable or require some level of human judgment. Indeed, the designer should have a global vision of all possible variants of the routines to define the appropriate behaviours of the SW robots, which becomes complicated when the number of variants increases. The issue is that in case where the flowchart diagram does not contain a suitable response for a specific situation, e.g., because of an inaccurate modeling activity, then the associated RPA scripts would not properly reflect the behaviour of the potential routine variant, forcing SW robots to escalate to a human supervisor at run-time, in contrast with the RPA philosophy.
   Objective: Once the routines to be automated and the user actions that constitute them (i.e., the routine-based logs) have been identified, the target is to automatically generate the flowchart diagrams (or/and the executable RPA scripts) describing the behaviours of the SW robots required to successfully execute the routines, rather than manually specify their conceptual and technical structure by means of interviews, walkthroughs and direct observation of workers (cf. steps 3 and 4).

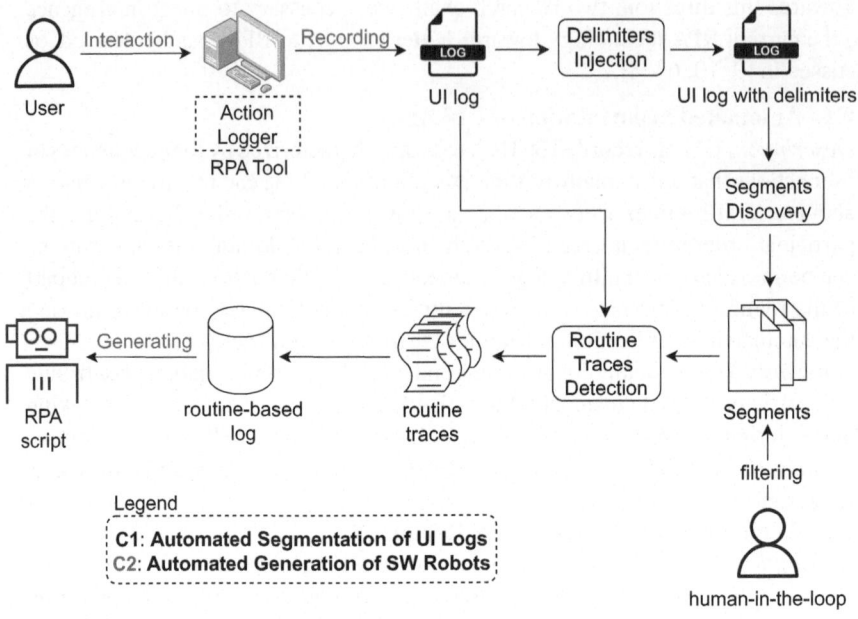

Fig. 1.1 Overview of the envisioned approach required to tackle **C1** and **C2**

To mitigate the involvement of skilled human experts in steps 1, 3, and 4 by tackling **C1** and **C2**, the thesis underlying this book is targeted to: *(i)* automatically understand which user actions contribute to which routines inside a UI log and *(ii)* automatically generate executable RPA scripts directly from the UI logs that record the user interactions with the SW applications involved in a routine execution, thus skipping the (manual) modeling activity of the flowchart diagrams.

To achieve these goals, as shown in Fig. 1.1, starting from an unsegmented UI log previously recorded with an RPA tool, the first stage of this research is to inject into the UI log the *end-delimiters* of the routines under examination. An end-delimiter is a dummy action added to the UI log immediately after the user action that is known to complete a routine execution. The knowledge of such end-delimiters is crucial to make the approach works, as discussed later in the book.

The second step of the approach is to automatically discover the most frequent observed segments of the routines as recorded into the UI log with the end-delimiters. *Routine segments* describe the different behaviours of the routine(s) under analysis in terms repeated patterns of performed user actions. An unsegmented UI log consists of sequential data of user actions performed on the UI of a computer system during many routines' executions. To achieve the segmentation task, we rely on a *frequent-pattern identification technique* [27] (which has been properly customized for this purpose) to automatically discover the observed segments of the routines as recorded into the UI log. In this phase, the risk exists that some wrong segments are discovered, i.e., not allowed from the real-world routines known to be valid at the outset.

To address this issue, a *human-in-the-loop interaction* that enables human experts to visualize the *declarative constraints* inferred by the discovered routine segments is required. Such rules describe the temporally extended relations between user actions that must be satisfied throughout a routine segment (e.g., an action a_1 must be eventually followed by an action a_2). In a nutshell, they collectively determine the observed behaviours of the routine segments from the UI log. This knowledge allows human experts to identify and remove those constraints that should not be compliant with any real-world routine behaviour, thus filtering out the not valid (i.e., wrongly discovered) routine segments.

Finally, starting from any of the remaining (valid) routine segments, a customized version of a *trace alignment* technique in Process Mining [2, 30] is employed to automatically detect and extract the *routine traces* by the original UI log. A routine trace represents an execution instance of a routine within a UI log. By identifying the routine traces, it is also possible to filter out those actions in the UI log that are not part of the routine under observation and hence are redundant or represent noise. Such traces are finally stored in a dedicated *routine-based log*, which captures precisely all the user actions that happened during many different executions of the routine, thus achieving the segmentation task (**C1**). Therefore, the outcome of our segmentation approach will be a collection of as many routine-based logs as are the number of valid routine segments.

The majority of state-of-the-art segmentation approaches can properly extract routine segments from unsegmented UI logs when the routine executions are not interleaved from each other. Only a few works are able to partially untangle un-segmented UI logs consisting of many interleaved routine executions, but with the assumption that any routine provides its own, separate universe of user actions. This is a relevant limitation since it is quite common that real-world routines may share the same user actions (e.g., copy and paste data across cells of a spreadsheet) to achieve their objectives. In this book, we propose a novel approach to the segmentation of UI logs [3] that aims to mitigate the issues as mentioned above, showing its ability to outperform existing literature approaches in terms of supported segmentation variants and to which extent the human-in-the-loop interaction is able to filter out the wrongly discovered routine segments. In particular, we evaluated the *robustness* of the frequent-pattern identification technique in (re)discovering the valid routine segments against synthetic UI logs of increasing complexity. Then, we also investigated the degree of *effectiveness*, *robustness*, and *usability* of the tool implementing the human-in-the-loop interaction step.

It is worth noticing that the commercial RPA tools can eventually employ routine-based logs to synthesize executable scripts in the form of SW robots that will emulate the routine behaviour on the UI without the manual modeling of the routines (**C2**). In this direction, a cross-platform software tool called SmartRPA[1][4, 5] was developed, which is able to generate executable RPA scripts, necessary to enact the SW robot

---

[1] SmartRPA is available at: https://github.com/bpm-diag/smartRPA

that implements the selected routine variant directly from a segmented UI log (i.e., a routine-based log). A *routine variant* is a specific execution of a routine that differs from the other executions (i.e., instances) of the same routine in at least one user action. Differently from the literature approaches to automated RPA scripts generation from UI logs, which enable to automate straightforward routines that have essentially no variance and do not require any human intervention, the SW robots generated by SmartRPA are obtained to handle the intermediate user inputs that are required during the routine execution, thus enabling to emulate the most suitable routine variant for any specific combination of user inputs as observed in the UI log. This makes the synthesis of SW robots performed by SmartRPA *reactive* to any user decision found during a routine execution. "Reactivity" highlights the fact that the behaviour of SW robots is determined immediately before their enactment, as it is driven by the specific user inputs required to execute the routine. This also means that reactivity enables the potential run-time generation of as many SW robots as the routine variants to be emulated.

SmartRPA has been validated on four non-functional requirements to measure the quality of the underlying approach. Specifically, we first perform many synthetic experiments employing UI logs of increasing complexity to assess the *robustness* and *feasibility* of SmartRPA to the identification of routine variants and variation points for the reactive synthesis of SW robots. A *variation point* is a point in the routine execution where a user choice needs to be made between multiple possible routine variants. Then, we performed a controlled experiment involving real users exploiting our RPA use case to investigate the *effectiveness* of the SmartRPA approach [6] when compared to a traditional model-based approach for the generation of SW robots. Finally, we quantify the *usability* of the UI provided by the tool implementing the SmartRPA approach.

While this Chapter serves as the basis for summarizing the performed author's research activities, the rest of the book is organized as follows. Chapter 2 presents the relevant background and preliminary concepts integrated with a real-life RPA use case useful to explain the proposed approaches to tackle **C1** and **C2**. Chapter 3 discusses the related work solutions to tackle the segmentation issue. Then, starting from the related work analysis, we derived an interactive approach to the automated segmentation of UI logs [3], which relies on: *(i)* a frequent-pattern identification technique (customized on a ad-hoc basis) to automatically derive the routine segments as recorded into a UI log (cf. Chapter 4), *(ii)* a human-in-the-loop interaction to filter out those segments not allowed (i.e., wrongly discovered from the UI log) by any real-world routine under analysis (cf. Chapter 5), and *(iii)* a routine traces detection component that leverages trace alignment in Process Mining to cluster all those user actions belonging to a specific segment into routine traces (cf. Chapter 6). The routine traces detection component can also be used as a supervised segmentation technique [11]. The *supervised* assumption, which consists of knowing a priori the structure of routines, may ease the segmentation task. Still, as a side effect, it may strongly constrain the discovery of routine traces only to the "paths" allowed by the routines' structure, thus neglecting that some valid yet infrequent routine variants may exist in the UI log. For this reason, the novelty of the proposed approach to the segmentation of UI logs [3] is to semi-automatically discover such structures in the form of routine segments and then use them as input for the supervised segmentation technique [11]. Then, Chapter 7 focuses on the design of the SmartRPA approach, presenting an algorithm to the automated identification of the variation points from many routine executions, to enable the selection of the most suitable routine variants to be implemented with a SW robot. Chapter 8 analyzes the architecture and the technical aspects of the tool implementing SmartRPA, describing also how the generated scripts can be automatically encoded in a format readable by the commercial RPA tool UiPath. Finally, Chapter 9 draws conclusions and traces future works.

Chapter 2
Background

This chapter provides all the prerequisites to understand the context and the scope of the book. First, in Section 2.1, we provide an overview of the BPM discipline and how BPs can be automated in this field. Next, in Section 2.2, we focus on a specific automation technology, called RPA, which aims at automating the routine tasks within the processes via executable scripts, while in Section 2.3 we provide the basic concepts of Process Mining. Then, in the remaining sections, we present some preliminary concepts used throughout the book. Section 2.4 outlines the definition of routines and SW robots while Section 2.5 places routines within the spectrum of BPs. Section 2.6 introduces a real-life RPA use case to illustrate the relevance of the research challenges being investigated. Section 2.7 describes the Petri net modeling language, which will be used to formally specify the interaction models required to represent the structure of the routines of interest explained in case study, while Section 2.8 introduces the notion of UI log. Then Section 2.9 provides the relevant background on trace alignment between UI logs and interaction models represented as Petri nets. Finally, Section 2.10 reviews the DECLARE modeling language used to describe a set of (temporally extended) constraints that must be satisfied throughout a routine segment.

2.1 Business Process Management

The growing interest of business organizations in understanding and improving their processes gave rise to specialized methods in process analysis, assessment, and refinement. Collectively, these methods can be framed under the BPM discipline. BPM aims at optimizing BPs via their design, administration, enactment and analysis. It allows business organizations to gain an operational advantage, reduce the costs and execution time of the processes, as well as related risks and errors [36]. Fig. 2.1 shows the BPM lifecycle consisting of the following phases:

- *Process identification*. This phase starts with the formulation of the business problem and the identification of the processes within the organization that are

S. Agostinelli: *Generating Executable Robotic Process Automation Scripts from Unsegmented User Interface Logs*, LNBIP 522, pp. 11–29, 2024.
https://doi.org/10.1007/978-3-031-61368-5_2

relevant to the problem. Their interrelations are analyzed, resulting in a process architecture used to select the process(es) for the subsequent phases of the cycle.

- *Process discovery.* This phase aims at discovering the workflow model of the selected process. Since the model corresponds to the current way the process is executed, it is called the as-is process model. For future analysis, the model is usually annotated with performance data (e.g., duration time of activities, waiting time between activities).

- *Process analysis.* During this phase, the constructed as-is process model is analyzed. During the analysis, all the issues associated with the model are identified, documented, quantified (if possible), and ranked with respect to their importance and potential impact.

- *Process redesign.* During this phase, the changes in the process required to address the documented issues are identified and evaluated. These changes are then applied to the as-is process model resulting in a new to-be process model that depicts how the process is expected to be executed.

- *Process implementation.* During this phase, the changes required to implement the to-be model are applied to the process. This may require structural changes (e.g., splitting, merging, or removing process activities, introducing parallelism) or process automation (e.g., developing and deploying IT systems).

- *Process monitoring.* During this phase, the newly implemented process is monitored and analyzed to verify whether it fully conforms to the intended execution. The presence of any deviations or errors may require another execution of the lifecycle.[1]

The process automation in BPM is achieved by developing and configuring IT systems that execute and coordinate the routine tasks within the process. Such systems aim at supporting the process participants (e.g., workers, or applications) in the execution of the process. This includes assigning tasks to process participants, providing them with the information required to perform the tasks, and monitoring the execution of the tasks [36].

A typical example of such a system is a BPMS. A BPMS is a system that supports design, analysis, execution, and monitoring of BPs based on their process models. The purpose of a BPMS is to coordinate an automated business process by assigning the process tasks to responsible resources at the required points in time.

The process model constructed at the process redesign phase of the BPM lifecycle can be used as a basis for the BPMS to coordinate the underlying process execution. However, this model is not executable as it depicts the process at a conceptual level. The information required to execute such a model is missing as it is not relevant for process analysis. In addition, the model does not include the instructions on how to react in the case of unexpected issues and exceptions that may be observed during the execution of the process, capturing only the "happy path".

Accordingly, to be deployed into a BPMS, the conceptual process model must be extended and transformed into an executable model. This involves the identification

---

[1] In some cases, the lifecycle is required to be repeated due to some external factors, e.g., the introduction of new policies, market situation changes, advancements in the technology.

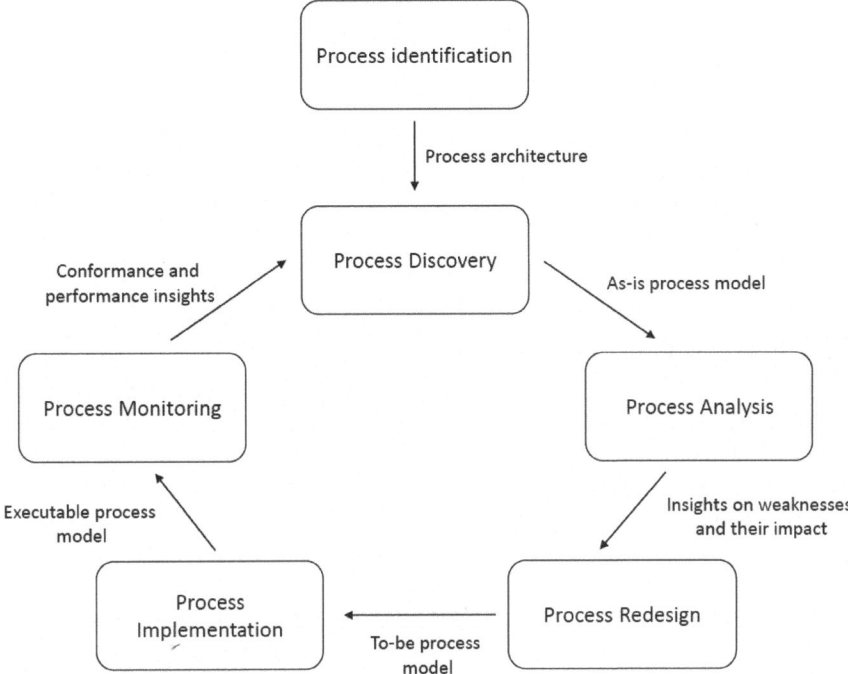

Fig. 2.1 The BPM Lifecycle

of the automation boundaries, the detection of the automatable and manual tasks within the process, and the refinement of the process model by introducing exception handlings, by specifying the data objects required in each step of the process, by defining the decision-making logic within the process, by decomposing and aggregating some of the process tasks, and by specifying their execution properties (e.g., data types, participant assignment rules).

The process automation with BPMS improves the quality and the efficiency of the BPs. A BPMS is able to reduce the workload by automatically assigning the tasks to the process participants and gathering all the relevant information needed for their execution. It enables organizations to become more flexible in managing and updating their BPs and applications and allows for the creation of a unified IT infrastructure by integrating all the applications within the organizations. Another advantage of automation with BPMS is the execution transparency, which leads to improvements in organizational control, as the organizations get better knowledge about their processes and resources. Finally, a BPMS ensures that the processes are executed as designed. It assures that the processes are performed in the best way possible and guarantees their compliance with the established policies and norms.

The main disadvantage of this type of automation is that it usually demands massive changes in the organizations. The existing IT systems have to be entirely redesigned or reconfigured to be integrated with a BPMS, as they are usually not designed to be process-aware. Therefore, the costs and benefits of automating certain processes must be thoroughly analyzed before applying such heavyweight automation.

Alternatively, the processes can be automated via RPA. RPA works at the level of user interfaces and does not require a lot of integration effort, thus allowing automation at a relatively low cost. In the next section, we overview RPA and show how it is different from traditional business process automation.

2.2 Robotic Process Automation

While many overlapping definitions of RPA can be found in the research literature, we adopt the one proposed by Gartner in 2017 [96], which defines RPA as *a class of tools that enable users to specify routines involving [if, then, else] statements on structured data, rules, user interface interactions, and operations accessible via APIs. Such routines are encoded as scripts that are executed by SW robots, operated via control dashboards.*

In the light of above, we can classify RPA as a specific type of process automation technology – a broader class of software tools that include BPMS, document workflow systems, and other types of workflow automation tools [36]. A key difference between RPA on the one hand and BPMSs, and workflow systems on the other is that RPA is meant to automate deterministic routines that involve automated steps where an interaction is performed with the user interface of an application. In contrast, BPMS and workflow systems are designed to automate processes that involve combinations of automated and manual tasks. Related to this distinction, BPMS, and workflow systems are designed to automate end-to-end processes consisting of multiple tasks performed by multiple types of participants (e.g., roles, groups). Meanwhile, RPA tools are developed to automate smaller routines, which correspond to individual tasks in a process, or even steps within a task, such as creating an invoice or a student record in an information system. As such, RPA tools and BPMSs are complementary, e.g., a BPMS may trigger an RPA tool to perform a given step in a process. Examples of RPA tools include Automation Anywhere,[2] Blue Prism[3] and UiPath.[4] The complete list of RPA tools can be inspected at [13] and to assess the current state of RPA landscape, we refer to the Gartner's Magic Quadrant.[5]

---

[2] https://www.automationanywhere.com/

[3] https://www.blueprism.com/

[4] https://www.uipath.com/

[5]   https://www.uipath.com/resources/automation-analyst-reports/
gartner-magic-quadrant-robotic-process-automation

2.3 Process Mining

The Process Mining [102] discipline bridges the gap between traditional model-based process analysis (simulation and other BPM techniques) and data-centric analysis techniques such as machine learning and data mining. Specifically, Process Mining is a family of approaches that extract information about BPs from process execution data, i.e., the *event logs*, recorded by standard enterprise systems available in organizations such as CRM (Customer Relationship Management) and ERP (Enterprise Resource Planning) systems. Each event in a log refers to an *activity* (i.e., a well-defined step in a BP) and is related to a particular *case* (i.e., a process instance). Events that belong to a case are ordered and constitute a single "run" of the process (often referred to as a *trace* of events). Event logs may store additional information about events such as *resources* (i.e., people and devices) executing or initiating the activities, *timestamps* indicating when the events occur, and data elements associated with the events. Data elements stored in the log can be either *event attributes*, i.e., data produced by the activities of a BP, or *case attributes*, namely data associated with the whole process instance. Table 2.1 shows an excerpt of an event log in the XES (eXtensible Event Stream) format [43].[6]

Table 2.1 Excerpt of an Event Log

Case ID	Timestamp	Activity	Resource	Cost
1	2024-03-14 10:00:02	register request	John	$50
1	2024-03-15 10:07:06	examine thoroughly	Jane	$100
1	2024-03-17 09:30:12	check ticket	Mark	$75
1	2024-03-18 12:03:18	decide	Emily	$60
1	2024-03-19 11:15:24	reject request	Sam	$60
2	2024-03-14 09:57:32	register request	Mark	$50
2	2024-03-14 11:09:12	check ticket	Mark	$75
2	2024-03-14 12:10:16	examine casually	John	$100
2	2024-03-17 11:05:22	decide	Emily	$60
2	2024-03-18 10:19:05	pay compensation	Simon	$60
...	...	...	...	...

In general, process mining techniques can be classified into three broad categories:

- *Process discovery*. Given an event log as input, process discovery aims to construct a model (e.g., a flowchart) that captures the activities performed during the process and their control flow relationships. A process model can be discovered in different formats, e.g., BPMN (Business Process Modeling Notation), EPC (Event-driven Process Chain), YAWL (Yet Another Workflow Language) models, Petri nets. The quality of the discovered model can be assessed according to four different

---

[6] In the last years, a new event logging format called OCEL (Object-Centric Event Logs) [42] has been proposed, where each event may refer to any number of objects (of different types) rather than a single case.

criteria [102]: fitness, precision, generalization, and simplicity. Some of these criteria are contradicting. For example, to achieve high precision, one has to sacrifice generalization, and vice versa. Balancing these quality metrics is the main challenge of process discovery.

- *Conformance checking.* Conformance checking aims at comparing existing process models against event logs that capture the same process, verify whether the model conforms to the real execution of the process, identify mismatches, and highlight the differences in behaviour.
- *Performance mining.* Performance mining analyzes the performance efficiency of the processes using the information from the event logs. The output of performance mining are performance statistics (e.g., throughput time, average activity execution time, waiting time), that can be used for process improvement, e.g., to identify the bottlenecks within the process.

The reader can refer to [101] for a 360 degree overview on Process Mining and the complete list of categorizations.

Example of major vendors of process mining tools are: several open-source tools (e.g., ProM, bupaR, PM4Py, and RapidProM) and over 40 commercial tools (e.g., Celonis, Disco/Fluxicon, Lana/Appian, Minit, Apromore, myInvenio/IBM, PAFnow, Signavio/SAP, and Timeline/Abby. As the time of writing, the Gartner's Magic Quadrant for process mining tools is: https://www.celonis.com/analyst-reports/gartner-magic-quadrant-2023/.

2.4 Routines and SW Robots

RPA moves around the concept of replacing routine work with automation. According to [31], a *routine* can be classified as a *structured process that reflects highly predictable and repetitive work with low flexibility requirements (i.e., the amount of variants to the expected process path is limited) and controlled interactions among process participants.*

As there is no unique definition of routines, we identify a key reference definition that, in our view, best represents the concept of routine in relation to the focus of this book. Depending on how the control dashboard is exploited, it is possible to distinguish among unattended and attended SW robots.

- *Unattended* SW robots are able to fully automate routines without any intermediate human intervention. This happens when all execution paths are always the same independently by the specific inputs provided to the routine executions. For example, for insurance claims management (when claims are received in a structured form), unattended SW robots offer an efficient solution for their automated processing and validation. However, any variant to the routine's expected behaviour is considered an exception and, thus, redirected to human supervision.
- *Attended* SW robots work alongside humans and are suitable for routines where some decisions or checks need to be made that require human judgment during

the routines' execution. Therefore, attended SW robots may require data from a user to properly progress the routine's enactment. For example, a document-driven routine lends itself to attended automation because a human is entering information via a document, and different values of the provided information may potentially trigger the execution of different variants of the routine. Let's consider the case of insurance claims rather than redirect routine variants to human supervision after the initial inspection in an insurance exception flow. An agent might feed claims to a SW robot that would collect different data points surrounding each claim. The SW robot would automatically validate and automate claims that fit the status quo, and return unusual claims to an agent for another level of review.

In a nutshell, unattended SW robots represent the simplest case of the attended perspective [63], since user inputs are not required for driving the routine's execution. On the other hand, attended SW robots are suitable in presence of *routine variants* recorded in the UI log. We define a routine variant as a *specific execution of a routine that differs from the other executions (i.e., instances) of the same routine by at least one event*. An *event* refers to the enactment of a user action (coupled with some execution data, like the name of the application where the action occurred, etc.) within a specific routine execution recorded in a UI log at a specific moment in time. The presence of different events in many routine executions *may potentially determine alternative behaviours* of the routine itself. This is particularly true when some events are triggered only by specific user inputs (and not by others) provided at the time of the routine execution. These events act as a *variation point* of the routine, where a user choice needs to be made between multiple possible variants. We will show an example of routine variants and variation points in Section 2.7.

2.5 RPA in the Spectrum of BPM

To better understand the types of processes that are best suited for RPA, a classification of BPs along a spectrum is presented in Fig. 2.2. The distinction among different types of BPs is made along the basis of the degree of structuring and predictability they exhibit, which directly influence the level of automation, control and support that can be provided, as well as the degree of flexibility that is required [31]. Along the spectrum, five different structuring levels can be identified:

- Structured.
- Structured with ad hoc exceptions.
- Unstructured with pre-defined segments.
- Loosely structured.
- Unstructured.

At the top of the spectrum, there are *Structured processes*, which are characterized by complete high predictability but the lowest level of flexibility. They can be described as a rigorously defined process with an end-to-end model that takes into

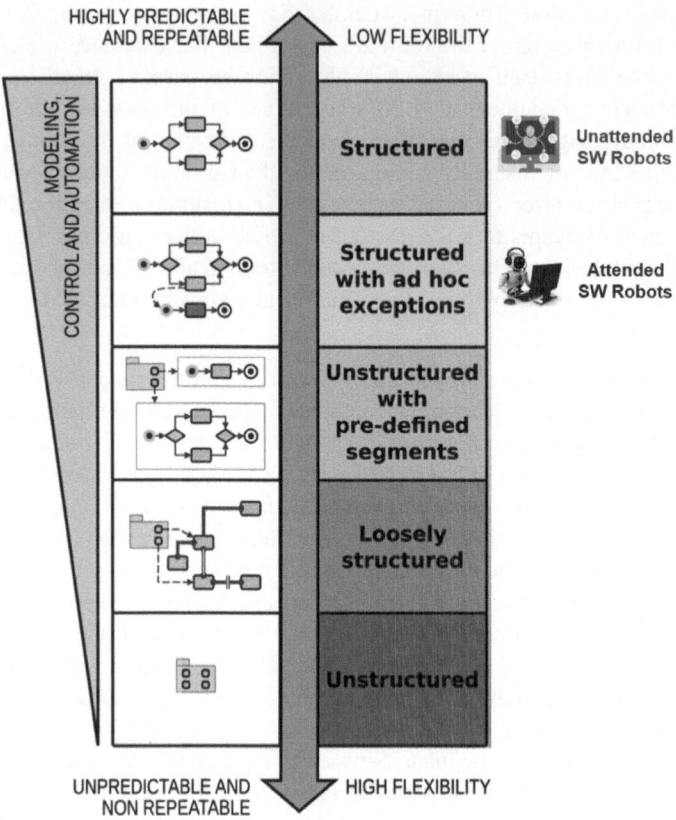

Fig. 2.2 The Spectrum of BPs

account all the process instance permutations. It means that all possible paths of the process are well-understood. Structured processes are usually very repetitive, including routine tasks and work that must be done regularly or at specified intervals. In addition, structured processes also feature low flexibility requirements, with no need to change or adjust, and controlled interactions among process participants. Typical examples of structured processes are production and administrative processes. In terms of the activities to be executed, their dependencies, and the resources performing the activities, the processing logic of this type of process is known in advance and can be predeterminate. As a consequence, all possible options and decisions that can be made during process enactment are captured in a process model, defined a priori. A process model is a representation of a BP consisting of a set of activities (a piece of work) and execution constraints between them, criteria to indicate the start and termination of the process, and information about participants, associated IT applications and data, etc. The distinctiveness of the process model of a structured process is that it can be repeatedly instantiated in a predictable and controlled manner. In this domain, the inputs and outputs are precise. Most process instances follow the same path. The process is expected to proceed through a high volume, producing millions of nearly identical outcomes with no exceptions.

The second class of BPs presented in the spectrum is the *Structured processes with ad hoc exceptions*. These processes share some characteristics with the structured processes, as they are both structured and reflect operational activities that typically comply with a predefined plan. Still, there is a significant difference in flexibility. The occurrence of external events and exceptions can make the process less rigid, thus requiring these processes to be adapted according to changing circumstances during the execution. Indeed, there is the possibility that the actual course of action may deviate from the predefined reference work practices, and this would consequently require process adaption strategies. In the presence of anticipated exceptions, possible deviations that can be encountered are predictable and defined in advance via exception handlers, typically pre-specified into the process model. The encountered exceptions are pretty predictable in advance, and the process can be modeled so that at every step, the user has the option of indicating that one of the prepared lists of exceptions occurred and some specific handling logic for each exception can be implemented. Contrarily, unanticipated exceptions can be only detected during the execution of a process instance, and their handling typically requires ad-hoc process changes at run-time.

In the middle of the spectrum, as the third presented BPs class, there are the *Unstructured processes with predefined segments* where work practices are somewhat unstructured and proceed on an ad-hoc basis. The overall process logic is not explicitly defined, but the existence of policies and regulations allows for identifying pre-definable, structured fragments. These fragments can refer to detailed, prescriptive procedures or may take the form of underspecified templates and guidelines (ambiguous because not sufficiently clear). Predefined process fragments need to be selected and adequately collected according to the features relating to each situation, case by case. Likewise, process parts that are undefined or uncertain can only be specified and incorporated in the range of the existing process model as the process evolves, and decisions regarding the specification of the process or parts of it have to be deferred.

The fourth class of BPs presented in the spectrum includes the *Loosely structured processes*, characterized by a higher level of flexibility and at the same time a lower level of predictability concerning the upper classes. The possible activities included in the process may be known and predefined, but their execution ordering is not entirely foreseeable as many possible execution alternatives are allowed. In particular, these processes can be made up of tasks that are not subject to fixed reference procedures. Nevertheless, there are constraints given by business rules that implicitly outline the scope of action of process participants, limiting their execution procedures. These constraints are used to describe processes defining the alternatives by prohibiting undesired execution behaviour.

Finally, at the bottom of the spectrum, the last class of BPs presented is the *Unstructured processes* characterized by the lowest level of predictability and contemporarily the highest level of flexibility. Furthermore, they can be stated that as unstructured, differently from a structured process where the frame is predeterminate, and the workflow must stick to it, here the structure of the process dynamically evolves with the process execution. Process participants, indeed, actively decide on the activities to be executed and their execution order based on their work knowledge and background. Knowledge workers rely on their experience to perform ad-hoc tasks on a case-by-case basis and handle unexpected changes in the operational context. For processes with these characteristics, only their goal is known a priori and reflect both workers know-how and collaboration activities driven by rules and events. No predefined models can be specified, and little automation can be provided.

An important element of the spectrum is the classification method presented on the left of Fig. 2.2: the triangle labelled "*modelling, control and automation*" depicts the direction towards automation potential increase. It is clear that the more the process is structured, the more it can be automated, and the less human intervention is needed. In other words, the higher the process is predictable and repeatable, the higher it can be automated, given its clear structure and predetermined process execution. On the other hand, the higher flexibility of a process you have, the less chance to be automated due to the process adaptability to different situations and the need to manage with non-predictable exceptions.

Based on this interpretation, we can position routines between the spectrum of *Structured processes* and *Structured processes with ad hoc-exceptions*, thus making the processes best suited for RPA the ones that meet the following characteristics:

- *Rule-based*: the logic of its workflow is defined utilizing if-then-else constructs.
- *Well-structured*: all possible execution paths are defined at design time, and exceptions and deviations are known and predictable.
- *Repetitive*: the execution flow is highly repeatable with low flexibility requirement.

This is exactly the case of the RPA use case explained in the following section, where the decision logic of the routines examined is expressed in terms of business constructs such as loops, parallel, and alternative branches, thus characterized by a low-level of flexibility (in terms of alternative behaviours to execute the routines under examination) and a high-level of predictability.

2.6 An RPA Use Case

In this section, we describe an RPA use case inspired by a real-life scenario at the Department of Computer, Control and Management Engineering (DIAG) of Sapienza Università di Roma. The scenario concerns the filling the travel authorization request form made by professors, researchers, and PhD students of DIAG for travel requiring prior approval. The request applicant must fill a well-structured Excel spreadsheet (cf. Fig. 2.3(a)) providing some personal information, such as her/his bio-data and the email address, together with further information related to the travel, including the destination, the starting/ending date/time, the means of transport to be used, the travel purpose, and the envisioned amount of travel expenses, associated with the possibility to request an anticipation of the expenses already incurred (e.g., to request

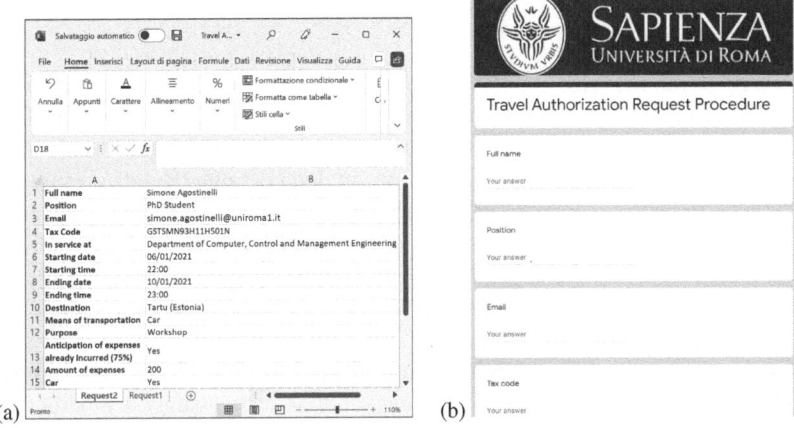

Fig. 2.3 UIs involved in the RPA use case

in advance a visa). When ready, the spreadsheet is sent via email to an employee of the Administration Office of DIAG, which is in charge of approving and elaborating the request. Concretely, for each row in the spreadsheet, the employee manually copies every cell in that row and pastes that into the corresponding text field in a dedicated Google form (cf. Fig. 2.3(b)), accessible just by the Administration staff. Once the data transfer for a given travel authorization request has been completed, the employee presses the "Submit" button to submit the data into an internal database.

In addition, if the request applicant declares that s/he would like to use her/his personal car as one of the means of transport for the travel, then s/he has to fill a dedicated (simple) web form required for activating a special insurance for the part of the travel that will be performed with the car. This further request will be delivered to the Administration staff via email, and the employee in charge of processing it can either approve or reject such request. In the end, the applicant will be automatically notified via email of the approval/rejection of the request.

The above procedure, which involves two main routines (in the following, we will denote them as R_1 and R_2), is performed manually by an employee of the Administration Office of DIAG, and it should be repeated for any new travel request. Routines such as these are good candidates to be encoded with executable scripts and enacted through a SW robot within a commercial RPA tool. However, unless there is complete a priori knowledge of the specific routines that are enacted on the UI and of their concrete composition (this may happen only if the exact sequence of user actions required to achieve the routines' targets on the UI is recorded in the context of controlled training sessions), their automated identification from an UI log is challenging, since the associated user actions may be scattered across the log, interleaved with other actions that are not part of the routine under analysis, and potentially shared by many routines.

2.7 Interaction Models as Petri Nets

The research literature is rich of notations for expressing human-computer dialogues as *interaction models*[7] that allow seeing at a glance the structure of user interactions with a UI [83, 33]. Existing notations can be categorized in two main classes: *diagrammatic* and *textual*. Diagrammatic notations include (among the others) various forms of state transition networks (STNs) [105], Petri nets [95], Harel state charts [45], flow charts [33], JSD diagrams [94] and ConcurTaskTrees (CTT) [81]. Textual notations include regular expressions [98], Linear Temporal Logic (LTL) [85], Communicating Sequential Processes (CSPs) [32], GOMS [53], modal action logic [24], BNF and production rules [38].

While there are major differences in expressive power between different notations, increased expressive power is not always desirable as it may suggest a harder to understand description, i.e., the dialogue of a UI can become unmanageable [33]. To guarantee a good trade-off between expressive power and understandability of the models, we decided to use *Petri nets* for their specification. Petri nets have proven to be adequate for defining interaction models [33, 82, 73]. They may contain exclusive choices, parallel branches and loops, allowing the representation of highly complex behaviours in a very compact way. Last but not least, Petri nets provide formal semantics, which helps to interpret the meaning of an interaction model unambiguously.

From a formal point of view, a Petri net $W = (P, T, S)$ is a directed graph with a set P of nodes called *places* and a set T of *transitions*. The nodes are connected via directed arcs $S \subseteq (P \times T) \cup (T \times P)$. Connections between two nodes of the same type are not allowed. Places are represented by circles and transitions by rectangles. Fig. 2.4 and Fig. 2.5 illustrate the Petri nets used to represent the interaction models of R_1 and R_2. Transitions are associated with *labels* reflecting the user actions (e.g., system commands executed, buttons clicked, etc.) required to accomplish a routine on the UI. For example, a proper execution of R_1 requires a path on the UI made by the following user actions:

- loginMail, to access the client email;
- accessMail, to access the specific email with the travel request;
- downloadAttachment, to download the Excel file including the travel request;
- openWorkbook, to open the Excel spreadsheet;
- openGoogleForm, to access the Google Form to be filled;
- getCell, to select the cell in the i-th row of the Excel spreadsheet;
- copy, to copy the content of the selected cell;
- clickTextField, to select the specific text field of the Google form where the content of the cell should be pasted;

---

[7] From now on, we refer to interaction models rather than process models since interaction models work at a lower level of abstraction, i.e., what actions have to be executed to complete a routine task. Conversely, process models focus on end-to-end BPs which are beyond the scope of this book since we do not aim to automate entire processes but instead routine tasks consisting of fine-grained user actions.

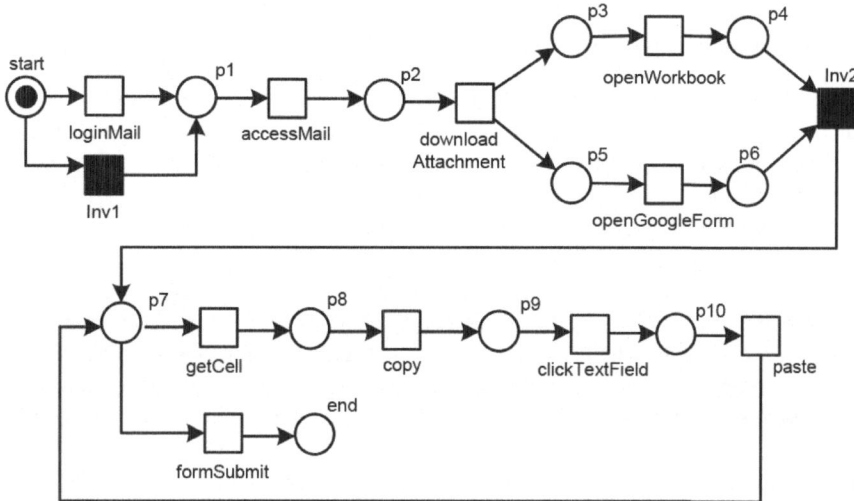

Fig. 2.4 Interaction model for R_1

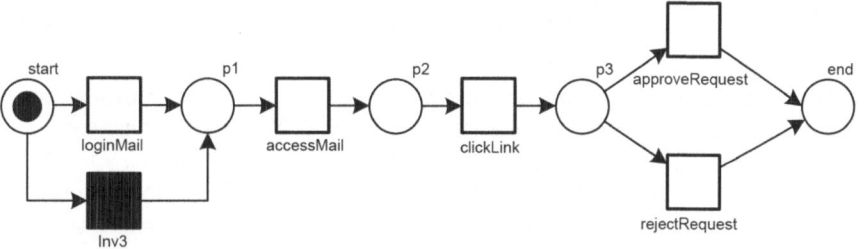

Fig. 2.5 Interaction model for R_2

- **paste**, to paste the content of the cell into the corresponding text field of the Google form;
- **formSubmit**, to press the button to finally submit the Google Form to the internal database.

Note that, as shown in Fig. 2.4, the user actions **openWorkbook** and **openGoogleForm** can be performed in any order. Moreover, the sequence of actions ⟨getCell, copy, clickTextField, paste⟩ will be repeated for any travel information to be moved from the Excel spreadsheet to the Google form. On the other hand, the path of user actions in the UI to properly enact R_2 is as follows:

- **loginMail**, to access the client email;
- **accessMail**, to access the specific email with the request for travel insurance;
- **clickLink**, to activate in the Google form the dialog box for approving or rejecting the car request;

- approveRequest, to press the button on the Google form that approves the request;
- rejectRequest, to press the button on the Google form that rejects the request.

Then, in the interaction models of R_1 and R_2, there are transitions that do not represent user actions but are needed to represent the structure of such models correctly. These transitions, drawn with a black-filled rectangle, are said to be "invisible", and are not recorded in the UI logs (cf. Inv1, Inv2 and Inv3).

Finally, in case of a car request to be evaluated (clickLink), the execution of approveRequest or rejectRequest is exclusive. Depending on the choice of the above user actions, two different variants of R_2 can be potentially emulated. However, the behaviour implied by R_2 semantically changes only after the enactment of the action clickLink, which requires an explicit user decision between the possibility of *accepting* or *rejecting* the personal car request. Therefore, the actions approveRequest and rejectRequest represent a *variation point* of the routine, that forks its execution flow into two well-distinguished exclusive branches.

To understand trace alignment in Process Mining (cf. Chapter 6), we also need to briefly illustrate the dynamic behaviour of a Petri net, i.e., its operational semantics. Given a transition $t \in T$, ${}^\bullet t$ is used to indicate the set of *input places* of t, which are the places p with a directed arc from p to t (i.e., such that $(p, t) \in S$). Similarly, $t^\bullet$ indicates the set of *output places*, namely the places p with a direct arc from t to p. At any time, a place can contain zero or more *tokens*, drawn as black dots. The state of a Petri net, i.e., its *marking*, is determined by the number of tokens in places. Therefore, a marking m is a function $m : P \rightarrow \mathbb{N}$. In any run of a Petri net, the number of tokens in places may change, i.e., the Petri net marking. A transition t is *enabled* at a marking m iff each input place contains at least one token, i.e., $\forall p \in {}^\bullet t$, $m(p) > 0$. A transition t can *fire* at a marking m if and only if it is enabled. As result of firing a transition t, one token is "consumed" from each input place and one is "produced" in each output place. This is denoted as $m \xrightarrow{t} m'$. In the remainder, given a sequence of transition firing $\sigma = \langle t_1, \ldots, t_n \rangle \in T^*$, $m_0 \xrightarrow{\sigma} m_n$ is used to indicate $m_0 \xrightarrow{t_1} m_1 \xrightarrow{t_2} \ldots \xrightarrow{t_n} m_n$, i.e., m_n is *reachable* from m_0.

Since the executions of a routine have a start and an end, the interaction models represented through Petri nets need to be associated with an initial and final marking. For example, in both routines of Fig. 2.4 and Fig. 2.5, the markings with respectively one token in place *start* or in place *end* are the initial and final marking (and no tokens in any other place). In the remainder of this book, we assume all Petri nets to be 1-bounded. A Petri net is 1-bounded if no place ever contains more than one token in any reachable marking from the initial marking. One-boundness is not a big limitation as the behaviour allowed by interaction models can be represented as 1-bounded Petri nets [33, 73].

2.8 User Interface Logs

In its raw form, a single UI log consists of a long sequence of user actions recorded during one user session.[8] Such actions include all the steps required to accomplish relevant routines using the UI of one or many SW application/s. For instance, in Fig. 2.6, we show a snapshot of a UI log captured using a dedicated action logger[9] (that we will discuss in Chapter 8) during the execution of R_1 and R_2.

	A	B	C	D	E	F	G	H	I	J
1	timestamp	user	category	application	event type	event src path	clipboard content	workbook	worksheet	cell content
2	2020-04-06 13:47	Simone	Mail	Outlook	loginMail					
3	2020-04-06 13:47	Simone	Mail	Outlook	accessMail					
4	2020-04-06 13:47	Simone	Mail	Outlook	downloadAttachment					
5	2020-04-06 13:47	Simone	MicrosoftOffice	Microsoft Excel	openWorkbook	C:\Users\Simone\Desktop\richiesta missione a	richiesta missione.xlsx	Foglio1		
6	2020-04-06 13:47	Simone	MicrosoftOffice	Microsoft Excel	openWindow	C:\Users\Simone\Desktop		richiesta missione.xlsx	Foglio1	
7	2020-04-06 13:47	Simone	MicrosoftOffice	Microsoft Excel	afterCalculate					
8	2020-04-06 13:47	Simone	MicrosoftOffice	Microsoft Excel	resizeWindow	C:\Users\Simone\Desktop		richiesta missione.xlsx	Foglio1	
9	2020-04-06 13:47	Simone	Browser	Chrome	openGoogleForm					
10	2020-04-06 13:47	Simone	MicrosoftOffice	Microsoft Excel	getCell			richiesta missione.xlsx	Foglio1	Simone Agostinelli
11	2020-04-06 13:47	Simone	Clipboard	Clipboard	copy		Simone Agostinelli			
12	2020-04-06 13:47	Simone	Browser	Chrome	clickTextField					
13	2020-04-06 13:48	Simone	Mail	Outlook	clickLink					
14	2020-04-06 13:48	Simone	Browser	Chrome	paste		Simone Agostinelli			
15	2020-04-06 13:48	Simone	Browser	Chrome	changeField					
16	2020-04-06 13:48	Simone	Browser	Chrome	approveRequest					
17	2020-04-06 13:48	Simone	MicrosoftOffice	Microsoft Excel	getCell			richiesta missione.xlsx	Foglio1	Dottorando
18	2020-04-06 13:48	Simone	Clipboard	Clipboard	copy		Dottorando			
19	2020-04-06 13:48	Simone	MicrosoftOffice	Microsoft Excel	resizeWindow	C:\Users\Simone\Desktop		richiesta missione.xlsx	Foglio1	
20	2020-04-06 13:48	Simone	Browser	Chrome	clickTextField					
21	2020-04-06 13:48	Simone	Browser	Chrome	paste		Dottorando			

Fig. 2.6 Snapshot of a UI log captured during the executions of R_1 and R_2

The employed action logger enables recording the *events* that happened on the UI, enriched with several data fields describing their "anatomy". For a given event, such fields are helpful to keep track of the name and the timestamp of the user action performed on the UI, the involved SW application or web page, the human/SW resource that performed the action, etc. There are similarities between UI logs and event logs used in process mining. Specifically, both types of logs consists of timestamped records, such that each record refers to the execution of an action (or activity in case of event logs) by a user. Also, each record may contain a payload consisting of one or more attribute-value pairs. Some commercial process mining vendors have exploited the similarities between UI logs and business process event logs to offer RPA-related features.

For the sake of understandability, we assume here that any user action associated to each event recorded in the UI log is mapped at most with one (and only one) Petri net transition, and that the collection of labels associated to the Petri net transitions is defined over the same alphabet as the user actions in the UI log,[10] i.e., the alphabet

---

[8] We interpret a user session as a group of interactions that a single user takes within a given time frame on the UI of a specific computer system.

[9] The dedicated action logger is integrated within SmartRPA, a self-developed RPA tool downloadable from this link: https://github.com/bpm-diag/smartRPA

[10] In [30], it is shown how these assumptions can be removed.

of user actions in the UI log is a *superset* of that used for defining the labels of Petri net transitions.

In the RPA use case, we can recognize in R_1 and R_2 a universe of user actions of interest $Z = \{A, B, C, D, E, F, G, H, I, L, M, N, O\}$, such that: A = loginMail, B = accessMail, C = downloadAttachment, D = openWorkbook, E = openGoogleForm, F = getCell, G = copy, H = clickTextField, I = paste, L = formSubmit, M = clickLink, N = approveRequest, O = rejectRequest.

As shown in Fig. 2.6, a UI log is not specifically recorded to capture pre-identified routines. A UI log may contain multiple and interleaved executions of one/many routine/s (cf. in Fig. 2.6 the blue/red boxes that group the user actions belonging to R_1 and R_2, respectively), as well as redundant behaviour and noise. We consider as *redundant* any user action that is unnecessary repeated during the execution of a routine, e.g., a text value that is first pasted in a wrong field by mistake and then is moved in the right place through a corrective action on the UI. On the other hand, we consider as *noise* all those user actions that do not contribute to the achievement of any routine target, e.g., a window that is resized. In Fig. 2.6, the sequences of user actions that are not surrounded by a blue/red box can be safely labelled as noise.

Based on the foregoing, our segmentation approach, whose main steps are extensively examined in chapters 4, 5 and 6 aims at identifying the most frequent observed routine segments from a UI log and then extracting all those user actions that match a distinguishable pattern as represented by the interaction model of a valid routine segment R (i.e., the output of the human-in-the-loop interaction step), thus filtering out redundant actions and noise. To be more specific, any sequence of user actions in the UI log that can be replayed from the initial to the final marking of the Petri net-based interaction model of R is said to be a *routine trace* of R, i.e., a complete execution instance of R within the UI log. For example, a valid routine trace of R_1 is $\langle A, B, C, D, E, F, G, H, I, L \rangle$. The interaction model of R_1 suggests that valid routine traces are also those ones where: *(i)* A is skipped (if the user is already logged in the client email); *(ii)* the pair of actions $\langle D, E \rangle$ is performed in reverse order; *(iii)* the sequence of actions $\langle F, G, H, I \rangle$ is executed several time before submitting the Google form. On the other hand, two main routine traces can be extracted from R_2: $\langle A, B, M, N \rangle$ and $\langle A, B, M, O \rangle$, again with the possibility to skip A, i.e., the access to the client email. Note that, within a routine trace, the concept of time is usually defined in a way that user actions in a trace are sorted according to the timestamp of their occurrence.

By analyzing the log, it can be noted that: A and B are shared by R_1 and R_2, as they are included in the interaction models of both routines. A is *potentially involved* in the enactment of any execution of R_1 and R_2, while B is *required by all* executions of R_1 and R_2, but it is not clear the association between the single executions of B and the routine executions they belong to. The complexity of the segmentation task here lies in understanding to which routine traces the execution of A and B belong.

We conclude this section by introducing the concept of *routine-based* log as a special container that stores all the routine traces extracted by a UI log and associated to a generic interaction model. Thus, the final outcome of the envisioned approach to the segmentation of UI logs will be a collection of as many routine-based logs as are the interaction models of interest.

2.9 Alignment between UI Logs and Interaction Models as Petri Nets

Trace alignment [2, 30, 29] is a conformance checking technique within Process Mining that is employed to replay the content of any trace of an event log against an interaction model represented as a Petri net, one event at a time. For each trace in the log, the technique identifies the closest corresponding trace that can be parsed by the model, i.e., an *alignment*, together with a *fitness* value, which quantifies how much the trace adheres to the interaction model. The fitness value can vary from 0 to 1. A value equals to 1 means a perfect matching between the trace and the model.

We perform trace alignment by constructing an alignment of a UI log U (note that we can consider the entire content of the UI log as a single trace) and an interaction model W (representing a valid routine segment) as a Petri Net, which allows us to exactly pinpoint where deviations occur. To this aim, the events in U need to be related to transitions in the model, and vice versa. Building this alignment is far from trivial, since the log may deviate from the model at an arbitrary number of places. To be more specific, we need to relate "moves" in the log to "moves" in the model in order to establish an alignment between an interaction model and a UI log. However, it may be that some of the moves in the log cannot be mimicked by the model and vice versa. We explicitly denote such "no moves" by $\gg$. In particular, we are interested in synchronous moves between U and W. If they exist, the user actions involved in such synchronous moves are extracted and stored into a routine-based log.

Definition 2.1 (Alignment Moves) Let $W = (P, T, S)$ be a Petri net and U be a UI log. A legal *alignment move* for W and U is represented by a pair $(q_U, q_W) \in (T \cup \{\gg\} \times T \cup \{\gg\}) \setminus \{(\gg, \gg)\}$ such that:

- (q_U, q_W) is a *move in log* if $q_U \neq \gg$ and $q_W = \gg$,
- (q_U, q_W) is a *move in model* if $q_U = \gg$ and $q_W \in T$,
- (q_U, q_W) is a *synchronous move* if $q_U = q_W$.

An alignment is a sequence of alignment moves:

Definition 2.2 (Alignment) Let $W = (P, T, S)$ be a Petri net with an initial marking and final marking denoted with m_i and m_f. Let also U be a UI log. Let Γ_W be the universe of all alignment moves for W and U. Sequence $\gamma \in \Gamma_W^*$ is an *alignment* of W and U if, ignoring all occurrences of $\gg$, the projection on the first element yields U and the projection on the second yields a sequence $\sigma'' \in T^*$ such that $m_i \xrightarrow{\sigma''} m_f$.

A move in log for a transition t indicates that t occurred when not allowed; a move in model for a visible transition t indicates that t did not occur, when, conversely, expected. Many alignments are possible for the same UI log and a Petri net.

$$\gamma_1 = \frac{\begin{array}{|c|c|c|c|}\hline A & B & M & N \\\hline A & B & M & N \\\hline\end{array}}{}$$

$$\gamma_2 = \frac{\begin{array}{|c|c|c|c|c|}\hline A & \gg & B & M & N \\\hline \gg & Inv3 & B & M & N \\\hline\end{array}}{}$$

Fig. 2.7 Alignments of $\langle A, B, M, N \rangle$ and the Petri net in Fig. 2.5.

For example, Fig. 2.7 shows two possible alignments for a UI log consisting of the following sequence of user actions $\langle A, B, M, N \rangle$ and the Petri net in Fig. 2.5, representing the interaction model of R_2. Note how moves are represented vertically. For example, as shown in Fig. 2.7, the first move of γ_1 is (A, A), i.e., a *synchronous move* of A, while the first and second move of γ_2 are a move in log and model, respectively. We aim at finding a complete alignment of U and W with minimal number of deviations (i.e., of moves in log/model) for visible transitions, also known in literature as *optimal alignments*. With reference to the alignments in Fig. 2.7, γ_1 have four synchronous moves and γ_2 have one move in log for visible transitions and one move in model for the invisible transition $Inv3$ (that does not count for the computation of the fitness value). As a consequence, γ_1 is an optimal alignment and can be returned. Note that its fitness value is exactly equal to 1, since it is consists only of synchronous moves enabling U to be completely replayed from the initial to the final marking of W. For the sake of simplicity, we are assuming here that all the deviations have the same severity. However, the severity of a deviation can be customized on a ad-hoc basis [30].

2.10 The DECLARE Modeling Language

In this section, we introduce the declarative process modeling language DECLARE [100] which is used in this book for expressing the declarative constraints that should be satisfied throughout the discovered routine segments from the UI log (cf. Chapter 5).

In general, instead of explicitly specifying the flow of the interactions among process activities, DECLARE describes a set of (temporally extended) constraints that must be satisfied throughout the process execution. The orderings of activities are implicitly specified by constraints and anything that does not violate them is possible during execution. Technically speaking, a DECLARE model $\mathcal{D} = (\mathcal{A}, \pi_\mathcal{D})$ consists of a set of possible activities $\mathcal{A}$ involved in a process and a collection of constraints π_d defined over such activities. DECLARE constraints are instantiations of templates, that is, patterns that define parameterized classes of properties. Templates have a graphical representation understandable to the user and analyst, but they also enjoy precise semantics in different logics [80] (e.g., LTL over finite traces), making them verifiable and executable.

In the following we summarize the existing DECLARE templates (the reader can refer to [100] for a full description of the language). DECLARE constraints can be divided into four main groups: existence, relation, mutual and negative constraints. Templates *existence(A)* and *absence(A)* require that A occurs at least once and never occurs in every process instance, respectively. Template *init(A)* specifies that A must occur in the first position of the process instance. Template *co-existence(A,B)* requires that if one of the activities *A* or *B* occur, the other one must also occur. Templates *choice* and *exclusive choice(A,B)* indicate that *A* or *B* occur eventually in each process instance. The exclusive choice template is more restrictive because it forbids *A* and *B* to occur both in the same process instance. The *responded existence(A,B)* template specifies that if *A* occurs, then *B* should also occur (either before or after *A*). The *response(A,B)* template specifies that when *A* occurs, then *B* should eventually occur after *A*. The *precedence(A,B)* template indicates that *B* can occur only if *A* has occurred before. Finally, the *succession(A,B)* template requires that both response and precedence relations hold between *A* and *B*. Templates *alternate response(A,B)* and *alternate precedence(A,B)* strengthen the response and precedence templates respectively by specifying that activities must alternate without repetitions in between. The *alternate succession(A,B)* template is the combination of the alternate response and alternate precedence templates. Even stronger ordering relations are specified by templates *chain response(A,B)* and *chain precedence(A,B)*. These templates require that the occurrences of *A* and *B* are next to each other. The *chain succession(A,B)* template is the combination of the chain response and chain precedence templates: *A* and *B* are always executed next to each other.

DECLARE also includes some negative constraints to explicitly forbid the execution of activities. The *not co-existence(A,B)* template indicates that *A* and *B* cannot occur together in the same process instance The *not responded existence(A,B)* template indicates that if *A* occurs in a process instance, *B* cannot occur in the same instance. According to the *not response(A,B)* template any occurrence of *A* cannot be eventually followed by *B*, whereas the *not precedence(A,B)* template requires that any occurrence of *B* is not preceded by *A*. The *not succession(A,B)* template requires that both not response and not precedence relations hold between *A* and *B*. Finally, according to the *not chain response(A,B)*, *not chain precedence(A,B)* and *not chain succession(A,B)* templates, *A* and *B* cannot occur one immediately after the other.

Part II
Automated Segmentation of UI Logs

Chapter 3
Segmentation in RPA

This book aims to explore the issue of automated segmentation in RPA (**C1**) through the evaluation of two research questions:

- **RQ1.1**: what are the variants of a segmentation solution needed to properly deal with different kinds of UI logs?
- **RQ1.2**: to what extent such variants are supported by literature approaches?

To answer these research questions, Section 3.1 first introduces the problem. Then, starting from the concrete RPA use case in the administrative sector (cf. Section 2.6) Section 3.2 explains how a segmentation technique should behave in presence of three different (and relevant) forms of UI logs, which may consist of: *(i)* several executions of the same routine, *(ii)* several executions of many routines without the possibility to have user actions in common, and *(iii)* several executions of many routines with the possibility to have user actions in common. Finally, Section 3.3 investigates how and if the current state-of-the-art segmentation approaches tackle such forms of UI logs.

3.1 The Segmentation Issue

Commercial RPA tools allow SW robots to automate a wide range of routines in a record-and-replay fashion and capture in dedicated UI logs the execution of high-volume routines previously performed by a human user on the interface of a computer system. As reported in [51], in the early stages of the RPA life-cycle it is required the support of skilled human experts to: *(i)* identify the candidate routines to automate by means of interviews and observation of workers conducting their daily work, *(ii)* record the interactions that take place during routines' enactment on the UI of software applications into dedicated UI logs, and *(iii)* manually specify their conceptual and technical structure (often in the form of flowchart diagrams) for defining the behaviour of SW robots.

S. Agostinelli: *Generating Executable Robotic Process Automation Scripts from Unsegmented User Interface Logs*, LNBIP 522, pp. 33–44, 2024.
https://doi.org/10.1007/978-3-031-61368-5_3

This approach is ineffective in the case of UI logs that keep track of many routines executions, since the designer should have a global vision of all possible variants of the routines to define the appropriate behaviours of SW robots, which becomes complicated when the number of variants increases. Indeed, in the presence of UI logs that collect information about several routines, the recorded actions are mixed in some order that reflects the particular order of their execution by the user, making the identification of candidate routines in a UI log a time-consuming and error-prone task. The issue to automatically understand which user actions contribute to a particular routine segment inside a UI log and cluster them into well-bounded *routine traces* (i.e., complete execution instances of a routine) is known as *segmentation* [9, 63].

A first approach proposed by Bosco et al. [22] makes this identification less time-consuming and error-prone, as it enables to automatically extract from a UI log, which records the UI interactions during a routine enactment, those routine steps to be automated with a SW robot. While this approach is effective in case of UI logs that keep track of single routine executions, i.e., there is an exact 1:1 mapping among a recorded user action and the specific routine it belongs to, it becomes inadequate when the UI log records information about several routines whose actions are mixed in some order that reflects the particular order of their execution by the user. In addition, since the same user action may belong to different routines, the automated identification of those user actions belonging to a specific routine is far from trivial.

Towards this direction, in their Robotic Process Mining framework [63], Leno et al. propose to exploit the User Interface (UI) logs recorded by RPA tools to automatically discover the candidate routines that can be later automated with SW robots. To date, when considering state-of-the-art RPA technology, it is evident that the RPA tools available in the market are not able to learn how to automate routines by only interpreting the user actions stored into UI logs [9]. The majority of state-of-the-art segmentation approaches can properly extract routine segments (i.e., repeated routine behaviours) from unsegmented UI logs when routines are not interleaved from each other. Only a few works are able to partially untangle unsegmented UI logs consisting of many interleaved routines, but with the assumption that any routine provides its own, separate universe of user actions. This is a relevant limitation since it is quite common that real-world routines may share the same user actions (e.g., copy and paste data across cells of a spreadsheet) to achieve their objectives.

To address the limitations as mentioned above, in this book, we have proposed an interactive approach to the discovery of routine traces from unsegmented UI logs [3], that is able to segment a UI log that records in an interleaved fashion many different routines with shared user actions but not the routine executions, thus losing in accuracy when there is the presence of interleaving executions of the same routine.

3.2 Identifying the Segmentation Variants

Given a UI log that consists of events including user actions with the same granularity[1] and potentially belonging to different routines, in the RPA domain *segmentation* is the task of clustering parts of the log together which belong to the same routine. In a nutshell, the challenge is to automatically understand which user actions contribute to which routines and organize such user actions in well-bounded routine traces [9, 63].

As shown in Section 2.8, in general, a UI log stores information about several routines enacted in an interleaved fashion, with the possibility that a specific user action is shared by different routines. Furthermore, actions providing redundant behaviour or not belonging to any of the routines under observation may be recorded in the log, generating noise that should be filtered out by a segmentation technique. Based on the above considerations, and on a concrete analysis of real UI logs recorded during the enactment of the routines presented in Section 2.6, i.e. R_1 and R_2, to address **RQ1.1** we have identified three main forms of UI logs, which can be categorized according to the fact that: *(i)* any user action in the log exclusively belongs to a specific routine (Case 1); *(ii)* the log records the execution of many routines that do not have any user action in common (Case 2); *(iii)* the log records the execution of many routines, and the possibility exists that some performed user actions are shared by many routines at the same time (Case 3). In the following, we analyze the three cases' characteristics and their variants. For the sake of understandability, we use a numerical subscript ij associated with any user action to indicate that it belongs to the $j - th$ execution of the $i - th$ routine under study. Of course, this information is not recorded in the UI log, and discovering it (i.e., identifying the subscripts) is one of the "implicit" effects of segmentation when routine traces are built.

Case 1. This is the case when a UI log captures many executions of the same routine. Of course, in this scenario it is impossible to distinguish between shared and non-shared user actions by different routines since the UI log keeps track only of executions associated with a single routine. Two main variants exist:

- **Case 1.1**. Starting from the use case in Section 2.6, let us consider the case of a UI log that records a sequence of user actions resulting from many non-interleaved executions of R_1 (cf. Fig. 3.1(a)). We also have the presence of some user actions that potentially belong *at the same time* to many executions of the routine itself. This is the case of loginMail, which can be performed exactly once at the beginning of a user session and can be "shared" by many executions of the same routine. Applying a segmentation technique to the above UI log would trivially produce a segmented UI log where the (already well-bounded) executions of R_1 are organized as different routine traces: the blue and grey vertical lines outline the routine traces, while the light blue line outlines R_1 itself.

- **Case 1.2**. The same segmented UI log is obtained when the executions of R_1 are recorded in an interleaved fashion in the original UI log (cf. Fig. 3.1(b)). Here,

---

[1] The UI logs created by generic action loggers usually consist of low-level events associated one-by-one to a recorded user action on the UI (e.g., mouse clicks, etc.).

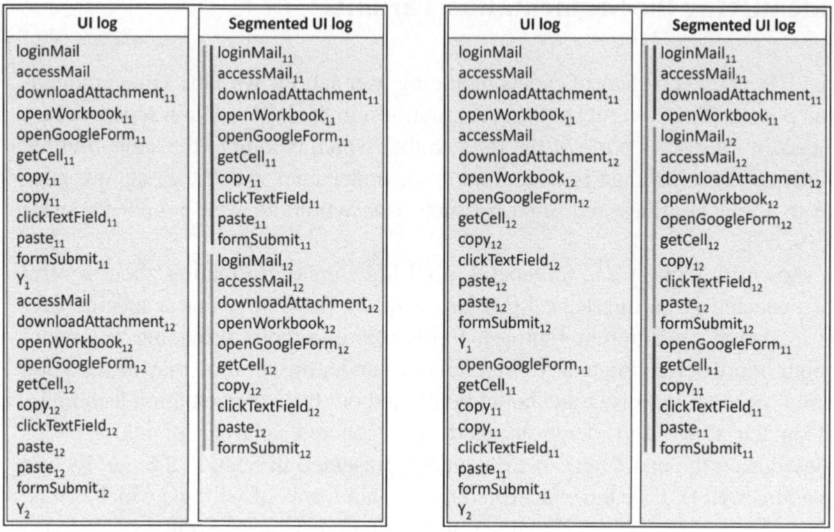

UI log	Segmented UI log
loginMail	loginMail$_{11}$
accessMail	accessMail$_{11}$
downloadAttachment$_{11}$	downloadAttachment$_{11}$
openWorkbook$_{11}$	openWorkbook$_{11}$
openGoogleForm$_{11}$	openGoogleForm$_{11}$
getCell$_{11}$	getCell$_{11}$
copy$_{11}$	copy$_{11}$
copy$_{11}$	clickTextField$_{11}$
clickTextField$_{11}$	paste$_{11}$
paste$_{11}$	formSubmit$_{11}$
formSubmit$_{11}$	loginMail$_{12}$
Y$_1$	accessMail$_{12}$
accessMail	downloadAttachment$_{12}$
downloadAttachment$_{12}$	openWorkbook$_{12}$
openWorkbook$_{12}$	openGoogleForm$_{12}$
openGoogleForm$_{12}$	getCell$_{12}$
getCell$_{12}$	copy$_{12}$
copy$_{12}$	clickTextField$_{12}$
clickTextField$_{12}$	paste$_{12}$
paste$_{12}$	formSubmit$_{12}$
paste$_{12}$	
formSubmit$_{12}$	
Y$_2$	

(a) Case 1.1

UI log	Segmented UI log
loginMail	loginMail$_{11}$
accessMail	accessMail$_{11}$
downloadAttachment$_{11}$	downloadAttachment$_{11}$
openWorkbook$_{11}$	openWorkbook$_{11}$
accessMail	loginMail$_{12}$
downloadAttachment$_{12}$	accessMail$_{12}$
openWorkbook$_{12}$	downloadAttachment$_{12}$
openGoogleForm$_{12}$	openWorkbook$_{12}$
getCell$_{12}$	openGoogleForm$_{12}$
copy$_{12}$	getCell$_{12}$
clickTextField$_{12}$	copy$_{12}$
paste$_{12}$	clickTextField$_{12}$
paste$_{12}$	paste$_{12}$
formSubmit$_{12}$	formSubmit$_{12}$
Y$_1$	openGoogleForm$_{11}$
openGoogleForm$_{11}$	getCell$_{11}$
getCell$_{11}$	copy$_{11}$
copy$_{11}$	clickTextField$_{11}$
copy$_{11}$	paste$_{11}$
clickTextField$_{11}$	formSubmit$_{11}$
paste$_{11}$	
formSubmit$_{11}$	
Y$_2$	

(b) Case 1.2

Fig. 3.1 Variants for Case 1

the segmentation task is more challenging because the user actions of different executions of the same routine are interleaved among each others, and it is not known a priori which execution they belong to.

Both variants of Case 1 are affected by *noise* or *redundant* actions. The logs contain elements of noise, i.e., user actions $Y_{k \in \{1,n\}} \in Z$ (remind that Z is the universe of user actions allowed by a UI log, as introduced in Section 2.8) that are not allowed by R_1, and redundant actions like copy and paste that are unnecessary repeated multiple times. Noise and redundant actions need to be filtered out during the segmentation task because they do not contribute to achieving the routine's target. In the following analysis, we do not consider the presence of noise and redundant actions anymore since their handling is similar for all the cases.

Case 2. In this case, a UI log captures many executions of different routines, assuming that the interaction models of such routines include only transitions associated with user actions that are exclusive for those routines. To comply with the latter constraint, let us suppose that in both interaction models of R_1 and R_2 the transitions loginMail and accessMail are not required. Four main variants of Case 2 can be identified:

- **Case 2.1**. Let us consider the UI log in Fig. 3.2(a). The output of the segmentation task would consist of a segmented log where the (already well-bounded) executions of R_1 and R_2 are organized as different routine traces: *(i)* the blue and grey vertical lines outline the routine traces of R_1, *(ii)* the yellow and orange vertical lines outline the routine traces of R_2, while *(iii)* the outer light blue and red

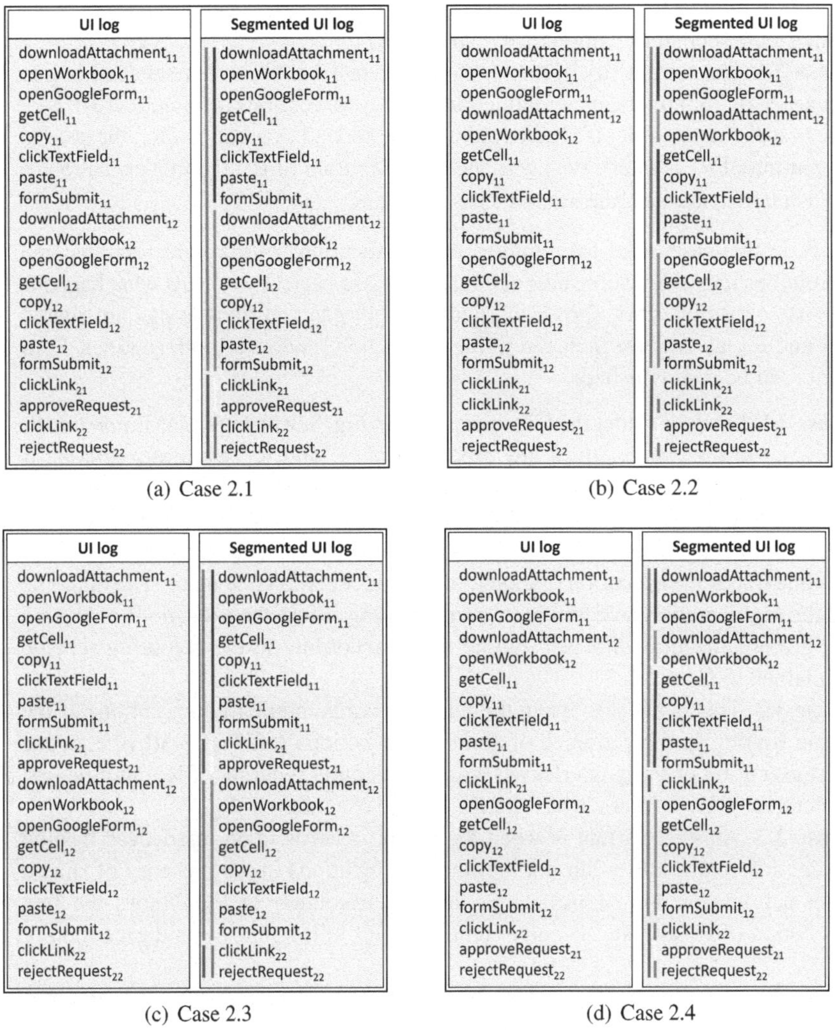

UI log	Segmented UI log
downloadAttachment$_{11}$	downloadAttachment$_{11}$
openWorkbook$_{11}$	openWorkbook$_{11}$
openGoogleForm$_{11}$	openGoogleForm$_{11}$
getCell$_{11}$	getCell$_{11}$
copy$_{11}$	copy$_{11}$
clickTextField$_{11}$	clickTextField$_{11}$
paste$_{11}$	paste$_{11}$
formSubmit$_{11}$	formSubmit$_{11}$
downloadAttachment$_{12}$	downloadAttachment$_{12}$
openWorkbook$_{12}$	openWorkbook$_{12}$
openGoogleForm$_{12}$	openGoogleForm$_{12}$
getCell$_{12}$	getCell$_{12}$
copy$_{12}$	copy$_{12}$
clickTextField$_{12}$	clickTextField$_{12}$
paste$_{12}$	paste$_{12}$
formSubmit$_{12}$	formSubmit$_{12}$
clickLink$_{21}$	clickLink$_{21}$
approveRequest$_{21}$	approveRequest$_{21}$
clickLink$_{22}$	clickLink$_{22}$
rejectRequest$_{22}$	rejectRequest$_{22}$

(a) Case 2.1

UI log	Segmented UI log
downloadAttachment$_{11}$	downloadAttachment$_{11}$
openWorkbook$_{11}$	openWorkbook$_{11}$
openGoogleForm$_{11}$	openGoogleForm$_{11}$
downloadAttachment$_{12}$	downloadAttachment$_{12}$
openWorkbook$_{12}$	openWorkbook$_{12}$
getCell$_{11}$	getCell$_{11}$
copy$_{11}$	copy$_{11}$
clickTextField$_{11}$	clickTextField$_{11}$
paste$_{11}$	paste$_{11}$
formSubmit$_{11}$	formSubmit$_{11}$
openGoogleForm$_{12}$	openGoogleForm$_{12}$
getCell$_{12}$	getCell$_{12}$
copy$_{12}$	copy$_{12}$
clickTextField$_{12}$	clickTextField$_{12}$
paste$_{12}$	paste$_{12}$
formSubmit$_{12}$	formSubmit$_{12}$
clickLink$_{21}$	clickLink$_{21}$
clickLink$_{22}$	clickLink$_{22}$
approveRequest$_{21}$	approveRequest$_{21}$
rejectRequest$_{22}$	rejectRequest$_{22}$

(b) Case 2.2

UI log	Segmented UI log
downloadAttachment$_{11}$	downloadAttachment$_{11}$
openWorkbook$_{11}$	openWorkbook$_{11}$
openGoogleForm$_{11}$	openGoogleForm$_{11}$
getCell$_{11}$	getCell$_{11}$
copy$_{11}$	copy$_{11}$
clickTextField$_{11}$	clickTextField$_{11}$
paste$_{11}$	paste$_{11}$
formSubmit$_{11}$	formSubmit$_{11}$
clickLink$_{21}$	clickLink$_{21}$
approveRequest$_{21}$	approveRequest$_{21}$
downloadAttachment$_{12}$	downloadAttachment$_{12}$
openWorkbook$_{12}$	openWorkbook$_{12}$
openGoogleForm$_{12}$	openGoogleForm$_{12}$
getCell$_{12}$	getCell$_{12}$
copy$_{12}$	copy$_{12}$
clickTextField$_{12}$	clickTextField$_{12}$
paste$_{12}$	paste$_{12}$
formSubmit$_{12}$	formSubmit$_{12}$
clickLink$_{22}$	clickLink$_{22}$
rejectRequest$_{22}$	rejectRequest$_{22}$

(c) Case 2.3

UI log	Segmented UI log
downloadAttachment$_{11}$	downloadAttachment$_{11}$
openWorkbook$_{11}$	openWorkbook$_{11}$
openGoogleForm$_{11}$	openGoogleForm$_{11}$
downloadAttachment$_{12}$	downloadAttachment$_{12}$
openWorkbook$_{12}$	openWorkbook$_{12}$
getCell$_{11}$	getCell$_{11}$
copy$_{11}$	copy$_{11}$
clickTextField$_{11}$	clickTextField$_{11}$
paste$_{11}$	paste$_{11}$
formSubmit$_{11}$	formSubmit$_{11}$
clickLink$_{21}$	clickLink$_{21}$
openGoogleForm$_{12}$	openGoogleForm$_{12}$
getCell$_{12}$	getCell$_{12}$
copy$_{12}$	copy$_{12}$
clickTextField$_{12}$	clickTextField$_{12}$
paste$_{12}$	paste$_{12}$
formSubmit$_{12}$	formSubmit$_{12}$
clickLink$_{22}$	clickLink$_{22}$
approveRequest$_{21}$	approveRequest$_{21}$
rejectRequest$_{22}$	rejectRequest$_{22}$

(d) Case 2.4

Fig. 3.2 Variants for Case 2

lines outline respectively the routines R_1 and R_2. In the following, the colouring scheme will be kept the same.

- **Case 2.2**. Similarly to what already seen in Case 1.2, many executions of the same routine may be interleaved among each other (cf. Fig. 3.2(b)), e.g., the first execution of R_1 is interleaved with the second execution of R_1, the first execution of R_2 is interleaved with the second execution of R_2, and so on.
- **Case 2.3**. Another variant is when the UI log records in an interleaved fashion many different routines but not the routine executions (cf. Fig. 3.2(c)), e.g., the

first execution of R_2 follows the first execution of R_1, the second execution of R_2 follows the second execution of R_1, and so on.

- **Case 2.4**. The complexity of the segmentation task becomes more challenging in presence of both interleaved routines and routine executions (cf. Fig. 3.2(d)), e.g., the first execution of R_1 is interleaved with the second execution of R_1, the second execution of R_1 is interleaved with the first execution of R_2, the first execution of R_2 is interleaved with the second execution of R_2.

Case 3. In this case, a UI log captures many executions of different routines, and such routines may share some user actions. This case perfectly reflects what happens in the use case of Section 2.6. In particular, loginMail and accessMail are shared by R_1 and R_2, as they are included in the interaction models of both routines. Four variants can be distinguished:

- **Case 3.1**. Let us consider the UI log depicted in Fig. 3.3(a). loginMail is *potentially involved* in the enactment of any execution of R_1 and R_2, while accessMail is *required by all* executions of R_1 and R_2, but it is not clear the association between the single executions of accessMail and the routine executions they belong to. The complexity of the segmentation task here lies in understanding to which routine traces the execution of loginMail and accessMail belong to. The outcome of the segmentation task will be a segmented log where the executions of R_1 and R_2 are organized as different routine traces according to the colouring scheme explained in Case 2.1.
- **Case 3.2**. This is the case when the UI log records interleaved executions of the same routine in the presence of shared user actions (cf. Fig. 3.3(b)), e.g., the first execution of R_1 is interleaved with the second execution of R_1, and the first execution of R_2 is interleaved with the second execution of R_2.
- **Case 3.3**. Another variant is when the UI log records in an interleaved fashion many different routines but not the routine executions in the presence of shared user actions (cf. Fig. 3.3(c)), e.g.: the first execution of R_2 follows the first execution of R_1 and the second execution of R_2 follows the second execution of R_1.
- **Case 3.4**. The segmentation task becomes more challenging in the presence of more complex UI logs consisting of both interleaved routines and routine executions with shared user actions (cf. Fig. 3.3(d)), e.g., the first execution of R_1 is interleaved with the second execution of R_1, the second execution of R_1 is interleaved with the first execution of R_2, and the first execution of R_2 is interleaved with the second execution of R_2.

The above three cases and their variants have in common that all the user actions are stored within a single UI log. It may happen that the same routine is spread across multiple UI logs, particularly when multiple users are involved in the execution of the routine on different computer systems. This case can be tackled by "merging" the UI logs where the routine execution is distributed into a single UI log, reducing the segmentation issue to one already analysed case. It is worth noticing that although the classification of cases and variants was illustrated with only two routines (interleaving or not), the classification is defined generically and applies to any number of routines.

UI log	Segmented UI log
loginMail	loginMail$_{11}$
accessMail	accessMail$_{11}$
downloadAttachment$_{11}$	downloadAttachment$_{11}$
openWorkbook$_{11}$	openWorkbook$_{11}$
openGoogleForm$_{11}$	openGoogleForm$_{11}$
getCell$_{11}$	getCell$_{11}$
copy$_{11}$	copy$_{11}$
clickTextField$_{11}$	clickTextField$_{11}$
paste$_{11}$	paste$_{11}$
formSubmit$_{11}$	formSubmit$_{11}$
accessMail	loginMail$_{12}$
downloadAttachment$_{12}$	accessMail$_{12}$
openWorkbook$_{12}$	downloadAttachment$_{12}$
openGoogleForm$_{12}$	openWorkbook$_{12}$
getCell$_{12}$	openGoogleForm$_{12}$
copy$_{12}$	getCell$_{12}$
clickTextField$_{12}$	copy$_{12}$
paste$_{12}$	clickTextField$_{12}$
formSubmit$_{12}$	paste$_{12}$
accessMail	formSubmit$_{12}$
clickLink$_{21}$	loginMail$_{21}$
approveRequest$_{21}$	accessMail$_{21}$
accessMail	clickLink$_{21}$
clickLink$_{22}$	approveRequest$_{21}$
rejectRequest$_{22}$	loginMail$_{22}$
	accessMail$_{22}$
	clickLink$_{22}$
	rejectRequest$_{22}$

(a) Case 3.1

UI log	Segmented UI log
loginMail	loginMail$_{11}$
accessMail	accessMail$_{11}$
downloadAttachment$_{11}$	downloadAttachment$_{11}$
openWorkbook$_{11}$	openWorkbook$_{11}$
openGoogleForm$_{11}$	openGoogleForm$_{11}$
accessMail	loginMail$_{12}$
openWorkbook$_{12}$	accessMail$_{12}$
openGoogleForm$_{12}$	openWorkbook$_{12}$
getCell$_{11}$	openGoogleForm$_{12}$
copy$_{11}$	getCell$_{11}$
clickTextField$_{11}$	copy$_{11}$
paste$_{11}$	clickTextField$_{11}$
formSubmit$_{11}$	paste$_{11}$
downloadAttachment$_{12}$	formSubmit$_{11}$
getCell$_{12}$	downloadAttachment$_{12}$
copy$_{12}$	getCell$_{12}$
clickTextField$_{12}$	copy$_{12}$
paste$_{12}$	clickTextField$_{12}$
formSubmit$_{12}$	paste$_{12}$
accessMail	formSubmit$_{12}$
accessMail	loginMail$_{21}$
clickLink$_{21}$	accessMail$_{21}$
approveRequest$_{21}$	loginMail$_{22}$
clickLink$_{22}$	accessMail$_{22}$
rejectRequest$_{22}$	clickLink$_{21}$
	approveRequest$_{21}$
	clickLink$_{22}$
	rejectRequest$_{22}$

(b) Case 3.2

UI log	Segmented UI log
loginMail	loginMail$_{11}$
accessMail	accessMail$_{11}$
downloadAttachment$_{11}$	downloadAttachment$_{11}$
openWorkbook$_{11}$	openWorkbook$_{11}$
openGoogleForm$_{11}$	openGoogleForm$_{11}$
getCell$_{11}$	getCell$_{11}$
copy$_{11}$	copy$_{11}$
clickTextField$_{11}$	clickTextField$_{11}$
paste$_{11}$	paste$_{11}$
formSubmit$_{11}$	formSubmit$_{11}$
accessMail	loginMail$_{21}$
clickLink$_{21}$	accessMail$_{21}$
approveRequest$_{21}$	clickLink$_{21}$
accessMail	approveRequest$_{21}$
downloadAttachment$_{12}$	loginMail$_{12}$
openWorkbook$_{12}$	accessMail$_{12}$
openGoogleForm$_{12}$	downloadAttachment$_{12}$
getCell$_{12}$	openWorkbook$_{12}$
copy$_{12}$	openGoogleForm$_{12}$
clickTextField$_{12}$	getCell$_{12}$
paste$_{12}$	copy$_{12}$
formSubmit$_{12}$	clickTextField$_{12}$
accessMail	paste$_{12}$
clickLink$_{22}$	formSubmit$_{12}$
rejectRequest$_{22}$	loginMail$_{22}$
	accessMail$_{22}$
	clickLink$_{22}$
	rejectRequest$_{22}$

(c) Case 3.3

UI log	Segmented UI log
loginMail	loginMail$_{11}$
accessMail	accessMail$_{11}$
downloadAttachment$_{11}$	downloadAttachment$_{11}$
openWorkbook$_{11}$	openWorkbook$_{11}$
openGoogleForm$_{11}$	openGoogleForm$_{11}$
accessMail	loginMail$_{12}$
openWorkbook$_{12}$	accessMail$_{12}$
openGoogleForm$_{12}$	openWorkbook$_{12}$
getCell$_{11}$	openGoogleForm$_{12}$
copy$_{11}$	getCell$_{11}$
clickTextField$_{11}$	copy$_{11}$
paste$_{11}$	clickTextField$_{11}$
formSubmit$_{11}$	paste$_{11}$
accessMail	formSubmit$_{11}$
downloadAttachment$_{12}$	loginMail$_{21}$
getCell$_{12}$	accessMail$_{21}$
copy$_{12}$	downloadAttachment$_{12}$
clickTextField$_{12}$	getCell$_{12}$
paste$_{12}$	copy$_{12}$
formSubmit$_{12}$	clickTextField$_{12}$
accessMail	paste$_{12}$
clickLink$_{21}$	formSubmit$_{12}$
approveRequest$_{21}$	loginMail$_{22}$
clickLink$_{22}$	accessMail$_{22}$
rejectRequest$_{22}$	clickLink$_{21}$
	approveRequest$_{21}$
	clickLink$_{22}$
	rejectRequest$_{22}$

(d) Case 3.4

Fig. 3.3 Variants for Case 3

Table 3.1 Literature approaches to tackle segmentation variants

Papers	Case 1		Case 2				Case 3			
	1.1	1.2	2.1	2.2	2.3	2.4	3.1	3.2	3.3	3.4
Abb et al. [1]	✓		✓		✓					
Agostinelli et al. [11]	✓	✓	✓	✓	✓	✓	~	~		
Agostinelli et al. [3]	✓		✓		✓		✓		✓	
Baier et al. [17]	✓		✓							
Bayomie et al. [18]	✓									
Bosco et al. [22]	✓		✓							
Kumar et al. [56]	✓		✓							
Leno et al. [60]	✓	~	✓	~	~	~				
Liu [68]	✓		✓	~	~	~				
Fazzinga et al. [37]	✓		✓				✓			
Ferreira et al. [39]	✓		✓							
Mannhardt et al. [70]	✓		✓							
Mărușter et al. [77]	✓									
Rebmann et al. [87]	✓		✓		✓		~		~	
Srivastava et al. [93]	✓		✓							
Urabe et al. [97]	✓	~	✓	~	✓	~				

3.3 State of the Art

In RPA, segmentation is still not so explored since the current practice adopted by commercial RPA tools for identifying the routine steps often consists of detailed observations of workers conducting their daily work. Such observations are then "converted" in explicit flowchart diagrams [51], which are manually modeled by expert RPA analysts to depict all the potential behaviours (i.e., the traces) of a specific routine. As the routine traces have already been (implicitly) identified in this setting, segmentation can be neglected.

On the other hand, following a similar trend that has been occurring in the BPM domain [72], the research on RPA is moving towards the application of intelligent techniques to automate all the steps of an RPA project, as proven by many recent works in this direction (see below). In this context, segmentation can be considered as one of the "hot" key research efforts to investigate [9, 63].

To answer **RQ1.2**, Table 3.1 summarizes the current literature techniques that could be leveraged to tackle the different variants of the segmentation issue. We will use ✓ to denote the full ability of an approach to deal with a specific UI log variant, while ~ denotes that the approach is only partially able to deal with a specific UI log variant (i.e., under certain conditions). In the following, we discuss to what extent existing literature approaches can support such variants. It is worth noticing that the assessment of the literature approaches is based on what was reported in the associated papers.

Concerning RPA-related techniques, Bosco et al. [22] provide a method that exploits rule mining and data transformation techniques, able to discover routines that are fully deterministic and thus amenable for automation directly from UI logs. This approach is effective in the case of UI logs that keep track of well-bounded routine executions (Case 1.1 and Case 2.1) and becomes inadequate when the UI log records information about several routines whose actions are potentially interleaved. In this direction, Leno et al. [60] propose a technique to identify execution traces of a specific routine relying on the automated synthesis of a control-flow graph, describing the observed directly-follow relations between the user actions. The technique in [60] is able to achieve cases 1.1, 1.2 and 2.1, and partially cases 2.2, 2.3 and 2.4, but (for the latter) it loses in accuracy in the presence of recurrent noise and interleaved routine executions. However, they are not able to handle UI logs that record in an interleaved fashion shared user actions of many different routines.

To tackle the main limitation of the above techniques, in this book we have presented an approach [3] that tackles the segmentation challenge relying on three main steps: *(i)* an ad-hoc frequent-pattern identification technique to automatically derive the observed routine segments from a UI log (cf. Chapter 4), *(ii)* a human-in-the-loop interaction to filter out those segments not allowed by any real-world routine execution (cf. Chapter 5), and *(iii)* a routine traces detection component that exploits trace alignment in Process Mining to cluster all user actions belonging to a specific routine segment into well-bounded routine traces (cf. Chapter 6). The approach is able to extract routine traces from unsegmented UI logs that record in an interleaved fashion many different routines but not the routine executions, thus losing in accuracy when there is the presence of interleaving executions of the same routine. In addition, it is also able to properly deal with shared user actions required by all routine executions in the UI log, thus achieving the cases 1.1, 2.1, 2.3, 3.1, and 3.3. It is worth noticing that the routine traces detection component can be employed as a supervised segmentation technique [11] able to achieve all variants of cases 1, 2, and (partially) 3, except when there are interleaved executions of shared user actions of many routines. In that case, the risk exists that a shared user action is associated with a wrong routine execution (i.e., Case 3.3 and Case 3.4 are not covered). While in [11], to make the technique works, it is required to know at the outset the structure (i.e., the interaction models) of the routines to identify in the UI log, in [3] this assumption has been mitigated by semi-automatically discovering such structures in the form of routine segments, and then used them as input for the routine traces detection component [11].

Even if more focused on traditional business processes in BPM rather than on RPA routines, Bayomie et al. [18] address the problem of correlating uncorrelated event logs in process mining in which they assume the model of the routine is known. Since event logs allow to store traces of one process model only, this technique is able to achieve Case 1.1 only. In the field of process discovery, Măruşter et al. [77] propose an empirical method for inducing rule sets from event logs containing the execution of one process only. Therefore, as in [18], this method is able to achieve Case 1.1 only, thus making the technique ineffective in the presence of interleaved and shared user actions. A more robust approach, developed by Fazzinga et al. [37], employs predefined behavioural models to establish which process activities belong to which process model. The technique works well when there are no interleaved user actions belonging to one or more routines since it cannot discriminate which event instance (but just the event type) belongs to which process model. This makes [37] effective to tackle Case 1.1, Case 2.1, and Case 3.1. Closely related to [37], there is the work of Liu [68]. The author proposes a probabilistic approach to learn workflow models from interleaved event logs, dealing with noises in the log data. Since each workflow is assigned with a disjoint set of operations, it means the proposed approach is able to achieve both cases 1.1 and 2.1, but partially cases 2.2, 2.3, and 2.4 (the approach can lose accuracy in assigning operations to workflows).

Differently from the previous works, Time-Aware Partitioning (TAP) techniques cut event logs based on the temporal distance between two events [93, 56]. The main limitation of TAP approaches is that they rely only on the time gap between events without considering any process/routine context. For this reason, such techniques cannot handle neither interleaved user actions of different routine executions nor interleaved user actions of different routines. As a consequence, TAP techniques are able to achieve cases 1.1 and 2.1.

There exist other approaches whose target is not to exactly resolve the segmentation issue. Many research works exist that analyze UI logs at different abstraction levels, which can be potentially valuable for realizing segmentation techniques. With the term *"abstraction"* we mean that groups of user actions to be interpreted as executions of high-level activities. Baier et al. [17] propose a method to find a global one-to-one mapping between the user actions that appear in the UI log and the high-level activities of a given interaction model. This method leverages constraint-satisfaction techniques to reduce the set of candidate mappings. Similarly, Ferreira et al. [39], starting from a state-machine model describing the routine of interest in terms of high-level activities, employ heuristic techniques to find a mapping from a "micro-sequence" of user actions to the "macro-sequence" of activities in the state-machine model. Finally, Mannhardt et al. [70] present a technique that maps low-level event types to multiple high-level activities (while the event instances, i.e., with a specific timestamp in the log, can be coupled with a single high-level activity). However, segmentation techniques in RPA must enable to associate low-level event instances (corresponding to user actions) to multiple routines, making abstractions techniques ineffective to tackle all those cases where is the presence of interleaving user actions of the same (or different) routine(s). Consequently, all abstraction techniques are effective to achieve Case 1.1 and Case 2.1 only.

The analysis of the related work has pointed out that the majority of literature approaches are able to properly extract routine traces from unsegmented UI logs when the routine executions are not interleaved from each others, which is far from being a realistic assumption. Only a few works [37, 11, 60, 68] have demonstrated the full or partial ability to untangle unsegmented UI logs consisting of many interleaved routine executions, but with any routine providing its own, separate universe of user actions. However, we did not find any literature work able to properly deal with user actions potentially shared by many routine executions in the UI log. This is a relevant limitation since it is quite common that a user interaction with the UI corresponds to the executions of many routine steps at once.

Moreover, it is worth noticing the majority of the literature works rely on the so-called *supervised* assumption, which consists of some a priori knowledge of the structure of routines. Of course, this knowledge may ease the task of segmenting a UI log. But, as a side effect, it may strongly constrain the discovery of routine traces only to the "paths" allowed by the routines' structure, thus neglecting that some valid yet infrequent routine variants may exist in the UI log. To this end, [97] proposes an unsupervised method that helps analyzers identify high-volume and repetitive operations from a UI log. The method is unsupervised, which means it does not require the analyzers to have knowledge of the tasks contained in the UI log or to set up any rules. This method segments the UI log into smaller sequences and clusters them into groups of tasks. Authors assumes that the same task generally consists of the same operations, and they analyze the co-occurrence of operations and segments the UI log into a sequence of operations under the same task by dividing the non-co-occurring area within the operations. Then, clustering is performed in accordance with the type of operations in the segmented sequences. The proposed technique robust for situations where the order of operations is varied among each task execution or when tasks are not executed from start to end due to interruptions (cf. Cases 2). However, authors do not take into account the possibility to have shared user actions among different tasks (cf. Cases 3).

It is worth to underline that process discovery techniques [15] can also play a relevant role in tackling the segmentation issue, as demonstrated by some literature works [68, 37, 18]. However, the problem is that most discovery techniques work with event logs containing behaviours related to the execution of a single process model only. And, more importantly, event logs are already segmented into traces, i.e., with clear starting and ending points that delimit any recorded process execution. Conversely, a UI log consists of a long sequence of user actions belonging to different routines without any clear starting/ending point. Thus, a UI log is more similar to a unique (long) trace consisting of thousands of fine-grained user actions. With a UI log as input, the application of traditional discovery algorithms seems unsuited to discover routine traces and associate them to some routine models. Even if more research is needed in this area, a first attempt is made up by [1] that proposes a technique to preprocess a UI log by means of trace clustering, which partitions it into smaller groups of similar traces. Then, it applies process discovery to those more cohesive sub-logs and it results in a set of less complex and better readable models, which jointly provide a better overview of the full log. However, these models are only well comprehensible in some cases (i.e., Case 1.1, Case 2.1 and Case 2.3), e.g., when the UI log does not contain neither interleaved routine executions nor shared user actions.

Finally, differently from the presented techniques that work in an offline manner, requiring user interaction data to be stored in event logs, [87] proposes an unsupervised task recognition from user interaction streams that works in an online setting. In the PhD Thesis' original form, [87] was previously neglected since at that time of writing it was not yet been developed. However, the technique assumes that routines are executed sequentially, i.e., one routine execution must be completed before another one is started, thus neglecting interleaving executions of different routines.

Chapter 4
Segments Discovery through Frequent-Pattern Identification

The results of the investigation conducted in the previous chapter, allow us to derive a new research question required to properly tackle **C1**:

- **RQ1.3**: Which steps are required to make the automated segmentation of UI logs less dependent by the intervention of RPA human experts?

To properly address **RQ1.3**, we envision an interactive approach to the automated segmentation of UI logs [3] that allows to automatically understand which user actions contribute to which routines inside a UI log and cluster them into well-bounded routine traces. To be more precise, as shown in Fig. 4.1, starting from an unsegmented UI log previously recorded by an RPA tool, the first step is to inject into the UI log the *end-delimiters* of the routines under examination. An end-delimiter is a dummy action added to the UI log immediately after the user action that is known to complete a routine execution. If we consider the case study of Section 2.6, an end-delimiter is always required after the final action of R_1, i.e., formSubmit, and after one of the final actions or R_2, i.e., approveRequest or rejectRequest. Here, we assumed that the knowledge of the final action(s) of a routine is given at the outset rather than to know a priori the structure (i.e., the interaction models) of the routines to identify in the UI log. Such information can be obtained, for example, by interviewing the users that are in charge of executing the routines of interest.

The second step of the approach automatically extracts the observed routines' behaviours (i.e., the routine segments) directly from the UI log with the end-delimiters. To this aim, we employ a *frequent-pattern identification technique* [27], which has been properly customized for this purpose.

Since from the previous step there is the possibility that some (not allowed) segments are identified as if they would be valid, the third step of the approach involves a *human-in-the-loop interaction* to filter out these segments (cf. Chapter 5). Specifically, we infer the declarative constraints (i.e., the temporally extended relations between user actions) that must be satisfied throughout a routine segment. In this way, we enable human experts to identify and remove those constraints that should not be compliant with any real-world routine behaviour, thus removing the wrongly discovered routine segments from the UI log.

© The Author(s), under exclusive license to Springer Nature Switzerland AG 2024
S. Agostinelli: *Generating Executable Robotic Process Automation Scripts from Unsegmented User Interface Logs*, LNBIP 522, pp. 45–49, 2024.
https://doi.org/10.1007/978-3-031-61368-5_4

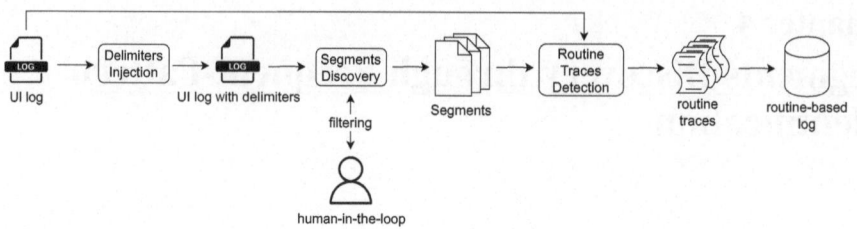

Fig. 4.1 Overview of our interactive approach to the segmentation of UI logs

Finally, starting from any of the remaining (valid) routine segments, we employ a customized version of a *trace alignment* technique in Process Mining [2], (cf. Algorithm 1, Chapter 6) to automatically detect and extract the *routine traces* by the original UI log. Such traces will be stored in a dedicated *routine-based log*. Therefore, the outcome of our segmentation approach will be a collection of as many routine-based logs as are the number of valid routine segments discovered by the approach itself. By identifying the routine traces, we are also able to filter out those actions in the UI log that are not part of the routine under observation and hence are redundant or represent noise. The implementation of our interactive approach to the segmentation of UI logs is available at: https://github.com/bpm-diag/AutSeg.

The overall approach can be considered as *semi-supervised*, since we know a priori the end-delimiters to be associated to any user action that ends a routine execution. On the other hand, the approach is not aware of the concrete behaviour of the routines of interest, which will be discovered by the approach itself, thus integrating the usage of automated techniques with the intervention of human experts in some specific points of the approach.

In the following section, we discuss in detail the frequent-pattern identification step (cf. Section 4.1), instantiating it over the RPA use case of Section 2.6. Then, we conclude the Chapter by showing its ability to outperform existing literature approaches in terms of supported segmentation variants (cf. Section 4.2).

4.1 Frequent-Pattern Identification

Pattern identification is a common task in data sequences analysis. As an example, in the field of smart spaces, patterns are identified in sensor logs representing human routines [66]. These patterns are then used to learn models of human behaviour that can be used at runtime for activity recognition or anomaly detection. In such a scenario, authors in [27] proposed an approach based on the minimum description length (MDL) principle. In this book, we have customized the technique presented in [27] for automatically identifying the routine segments from UI logs with the end-delimiters properly converted into ad-hoc datasets.

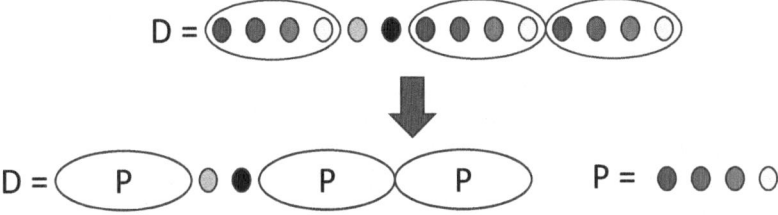

Fig. 4.2 A dataset compression step in segments discovery

The algorithm takes a dataset of a sequence of sensor events witnessing human interactions with the environment as input. At each step, the algorithm looks for patterns that best compress the dataset. A pattern consists of a specific sequence of sensor events and all of their occurrences in the dataset. In our RPA application scenario, the sensor events represent the user actions involved in each routine(s) execution(s), and the frequent patterns are the discovered routine segments.

Starting from a single pattern for each different sensor event, the algorithm at each step tries to extend patterns aiming at the best compression possible. Every instance of the pattern, in particular, is replaced by a symbol associated with the pattern. The compression of a dataset D given a pattern P is given by the formula $\frac{DL(D)}{DL(D|P)+DL(P)}$, where $DL(D)$ represents the description length, measured for example in bits of the dataset with the current patterns, $DL(D|P)$ represents the description length of D if all of the occurrences of P are replaced with a symbol, and $DL(P)$ represents the description length of the pattern, which must be taken into account in compression evaluation. The algorithm stops as soon as no further compression is possible, returning all the patterns found (i.e., all the discovered routine segments). Fig. 4.2 shows a compression step where a pattern P of repeating events (for simplicity, colours have been used instead of labels) is identified and the dataset is compressed accordingly. Noteworthy, for certain parts of the dataset, no pattern is found whose definition improves compression (except the initial patterns of length one).

We show now how an execution instance of the above algorithm can be applied to the following UI log (that already includes the end-delimiters) generated from the case study of Section 2.6: $U = \{A, B, C_{11}\ D_{11}, E_{11}, F_{11}, G_{11}, H_{11}, I_{11}, L_{11}, X, B, M_{21}, N_{21}, Z, B, C_{12}, D_{12}, E_{12}, F_{12}, G_{12}, H_{12}, I_{12}, L_{12}, X, B, M_{22}, O_{22}, Z, \ldots, A, B, C_{1(i-1)}, Y_1, D_{1(i-1)}, E_{1(i-1)}, F_{1(i-1)}, G_{1(i-1)}, G_{1(i-1)}, G_{1(i-1)}, H_{1(i-1)}, I_{1(i-1)}, L_{1(i-1)}, X, B, M_{2(i-1)}, N_{2(i-1)}, Z, B, Y_{n-1}, C_{1i}, D_{1i}, E_{1i}, Y_n, F_{1i}, G_{1i}, H_{1i}, I_{1i}, I_{1i}, I_{1i}, L_{1i}, X, B, M_{2i}, O_{2i}, Z\}$. For the sake of understandability, we use a numerical subscript ji associated with any user action to indicate that it belongs to the $i-th$ execution of the $j-th$ routine under study. This information is not recorded into the UI log, and discovering it (i.e., identifying the subscripts) is one of the "implicit" effects of segmentation when routine traces are built. Note that A and B are not decorated with subscripts since they can potentially belong to executions of R_1 or R_2. The log contains elements of noise, i.e., user actions $Y_{k \in \{1,n\}}$ that are not allowed by R_1 and R_2, and redundant actions like G and I that are unnecessary repeated multiple times. X and Z are the end-delimiters for the executions of R_1 and R_2.

The delimiters injection stage is crucial to drive the discovery of the largest possible set of valid routine segments. Otherwise, the technique would detect only a small subset of them. For example, let us suppose that the UI log includes only user actions related to two routines, A and B, without the presence of any end-delimiter. In this case, the UI log will likely include different sequences of consecutive routine segments of the kind A*, B* or AB*. In this condition, any compression algorithm will likely merge multiple routine segments into cumulative symbols (e.g., AAA, BB, ABAB) rather than highlighting single routine executions. This issue becomes less relevant when there are no repetitive actions between the execution of two separate routines. However, while the latter assumption is reasonable in recording human habits, it is far from being realistic in the case of UI logs recording low-level user actions performed during the interaction with a computer system.

Based on the foregoing, the output of the segments discovery stage is represented by a set of identified frequent segments (some of them may not be compliant with the real-world routine behaviours, see the next section), as follows:

- $\{\langle F, G \rangle, \langle C, D, E \rangle, \langle H, I, L \rangle, \langle C, D, E, F, G, H, I, L \rangle, \langle B, C, D, E, F, G, H, I, L \rangle, \langle A, B, C, D, E, F, G, H, I, L \rangle\}$
- $\{\langle A, B \rangle, \langle B, M \rangle, \langle B, M, O \rangle, \langle B, M, N \rangle\}$

4.2 Assessing the Robustness of the Segments Discovery Stage

In this section, we evaluate the *robustness* of our approach in the presence of UI logs of a growing size that provide an increasing amount of routine variants. Specifically, we assessed to what extent the approach is able to (re)discover routine segments that are known to be recorded into the input UI logs. We have synthetically generated 144 different UI logs in a way that each UI log consists of 1000 routine executions and is characterized by a unique configuration by varying the following inputs:

- *valid_routine_segments*: number of different routines segments (5/10/15/20), in terms of allowed behaviours, included in the UI log.
- *alphabet_size*: size of the alphabet of user actions for each segmentation case: Case 1 (13/18/23/28); Case 2 (15/16/18/21); Case 3 (13/15/17/20).
- *valid_traces*: percentage of allowed behaviours recorded into the UI log (50%/70%/100%). The remaining portion of the UI log (50%/30%) may be dirty, i.e., it contains routine executions potentially affected by noise.
- *percentage of noise* in the remaining (dirty) portion of the UI log (10%/20%).

The synthetic UI logs generated for the test and the complete list of results can be analyzed at: http://tinyurl.com/icsoc2021. Due to the long list of results we present in Table 4.1 only a view in one of the most complex cases to tackle. The results indicate that the approach scales very well in case of an increasing number of

Table 4.1 Experiments' results. For each segmentation case the number of actions is 28, 21 and 20 (resp.). Only logs with 20 different allowed segments are shown here, and the number of valid routine behaviours is the 70% of the 1000s that were introduced in the UI logs, while the other 30% may be affected by noise.

Case 1	# discovered segments (valid/wrong)		
Noise	0%	10%	20%
no repetitive actions	20/2	20/88	20/118
repetitive actions	20/11	16/161	16/179
Case 2	# discovered segments (valid/wrong)		
Noise	0%	10%	20%
no repetitive actions	20/2	20/59	20/69
repetitive actions	20/10	20/132	20/136
Case 3	# discovered segments (valid/wrong)		
Noise	0%	10%	20%
no repetitive actions	20/6	20/53	20/67
repetitive actions	20/13	20/146	20/170

different routine segments to be discovered and with an alphabet of user actions of growing size. The computation time is not shown since it ranges from milliseconds for UI logs with 5 different routine segments up to a few seconds for UI logs with 20 segments. This result was expected since more segments in a UI log means more executions to analyze and interpret.

By analyzing the results, we can infer that the approach is able to discover the same allowed routine segments that were synthetically introduced in the routine executions recorded in the UI logs, achieving the following segmentation cases: 1.1, 2.1, 2.3, 3.1 and 3.3. On the other hand, our approach seems to lack in the computation of valid routine segments in the presence of repetitive user actions (i.e., user actions that are repeated in a loop) when there are several routine segments generated by different executions of the same routine. This is because similar sequences of user actions tend to be compressed together, and since they are generated from the same routine, the risk exists that different sequences are wrongly recognized as the same and bounded together, thus leading to a number of routine segments lower than ones that were synthetically introduced.

Chapter 5
Human-in-the-loop Interaction through SCAN

Once the routine segments have been discovered, the possibility exists that many of them represent not allowed routine behaviours. This happens because a UI log combines the execution of several routines that are usually interleaved from each others. In addition, in case of routines that make use of the same kinds of user actions to achieve their goals, it may happen that new patterns of repeated user actions, which represent potential not allowed routine segments, are rather detected as valid ones within the UI log. Towards this direction we realized a software tool, called SCAN (Segments Compliance ANalisys), that concretely implement the human-in-the-loop interaction step allowing users to filter out those routines' segments not compliant with any real-world routine behaviours. In the following section, we discuss in detail the required steps to enact the human-in-the-loop interaction step through SCAN, instantiating it over the RPA use case of Section 2.6. Then, we measured the impact of the human-in-the-loop interaction to filter out the wrongly discovered routine segments. Specifically, we present the results of SCAN to investigate to which extent it satisfies three relevant non-functional requirements, namely *effectiveness* (cf. Section 5.2), *robustness* (cf. Section 5.3) and *usability* (cf. Section 5.4). The target is to understand if SCAN can potentially complement the traditional solutions provided by open-source Process Mining tools for helping users to perform the segmentation task in RPA.

5.1 Leveraging the human-in-the-loop

On the basis of the experiments performed in Section 4.2, it becomes clear that the employed frequent-pattern identification algorithm is able to (re)discover the allowed routine segments that are known to be recorded in the input UI logs. However, since there is the possibility that some (not allowed) segments are identified as if they would be valid, a *human-in-the-loop interaction* is required to filter out all those routine segments representing behaviours that should not be allowed by any real-world routine of interest.

S. Agostinelli: *Generating Executable Robotic Process Automation Scripts from Unsegmented User Interface Logs*, LNBIP 522, pp. 51–58, 2024.
https://doi.org/10.1007/978-3-031-61368-5_5

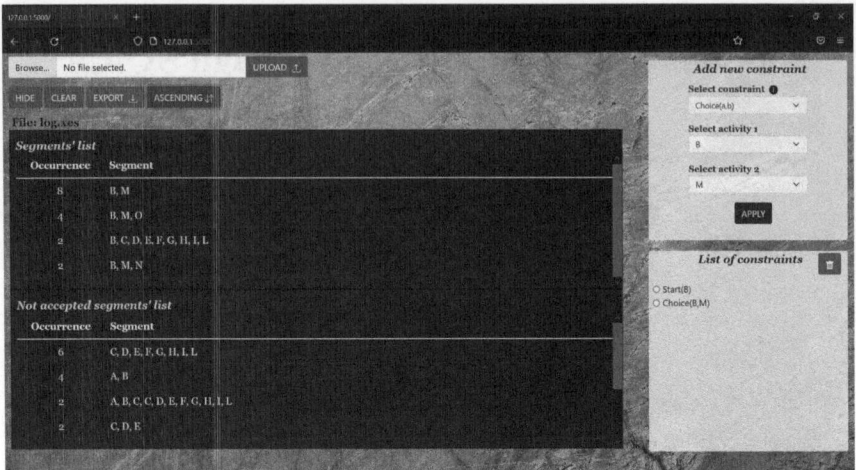

Fig. 5.1 GUI of SCAN

To address this issue, we developed a stand-alone web application called SCAN[1] (*Segments Compliance ANalysis*), which allows to support human experts in performing the *human-in-the-loop* step. The tool enables to visualize the declarative constraints (i.e., the temporally extended relations between user actions) that must be satisfied throughout the discovered routine segments from the UI log, as shown in Fig. 5.1. The constraints are represented using DECLARE, a well-known declarative process modeling language introduced in [100]. For a detailed summary on DECLARE, we let the readers refer to Section 2.10. This knowledge allows human experts to identify and remove those constraints that should not be compliant with any real-world routine behaviour. Detecting and removing these constraints means to filter out all the not allowed (i.e., wrongly discovered) routine segments. We notice that the use of declarative notations has been already demonstrated as an effective tool to visually support expert users in the analysis of event logs [89].

For example, if we consider the discovered segment $\langle C, D, E \rangle$ (cf. Section 4.1), the following (simple) DECLARE constraints (among the others) hold: *Init(C)* and *End(E)*, meaning that routines' executions starting with C or ending with E have been discovered into the UI log. An expert user that is aware of the behaviour of the real-world routines under analysis can immediately understand that the above DECLARE constraints should not hold in reality, since R_1 and R_2 can start only with **A** or **B** and end with L, O or N. For this reason, the above DECLARE constraints can be considered both as wrongly representative of the routines under analysis. As a consequence, all the discovered segments for which one of the above DECLARE constraints hold can be immediately discarded. For the sake of readability, we do not show here all the DECLARE constraints that hold for any of the discovered segments. However, we point out that the iterative analysis of the DECLARE constraints associated

---

[1] https://github.com/bpm-diag/SCAN

to the discovered segments will support the human experts to easily detect and filter
out those segments that must not be later emulated by SW robots. Starting from the
list of discovered routine segments of Section 4.1, the list of valid routine segments
for our case study outlined in Section 2.6 is the following:

- $W_1 = \langle A, B, C, D, E, F, G, H, I, L \rangle$
- $W_2 = \langle B, C, D, E, F, G, H, I, L \rangle$
- $W_3 = \langle B, M, O \rangle$
- $W_4 = \langle B, M, N \rangle$

5.2 Evaluating the Effectiveness of SCAN

An approach that simplifies the segmentation task in RPA, and in particular the
inspection of routine segments required to filter out the not allowed ones in the presence of many routine variants, can be considered as a relevant artefact to investigate.
Consequently, the research question (**RQ1.4**) we aim to investigate is the following
one: "*What is the effectiveness of employing an approach that semi-automatically
filters out the not allowed routine segments, thus neglecting the (manual) identification stage of the not allowed real-word routine behaviour, through declarative
constraints?*".

To address **RQ1.4** we enacted a controlled experiment involving a sample of 18
Master students of the course of Process Management and Mining (PMM) held at
Sapienza University of Rome, to investigate the effectiveness of employing SCAN to
perform the segmentation task when compared to DISCO[2]. Specifically, we selected
DISCO as target Process Mining tool since it provides user-friendly functionalities,
integrated with filtering facilities that allows to filter out the not allowed routine
segments as stored into event logs.

The user study was conducted as follows. Two case studies of increasing complexity were submitted to two different user groups of PMM students. The provided
case studies are inspired by the one presented in Section 2.6 and we refer to them
as Case Study #1 (cf. Fig. 2.5) and Case Study #2 (cf. Fig. 2.4). A first group of
9 PMM students were instructed to perform the case studies #1 and #2 exclusively
with DISCO. We denote with p_{A1} this first group of users. In parallel, a second
group of 9 PMM students received the same instructions of group p_{A1} but they are
asked to use SCAN rather than DISCO. We denote with p_{A2} this second group of
users. It is worth noticing that all the PMM students involved in the user study can
be considered expert users in business process modelling and automation.

To assess the effectiveness of SCAN in filtering out the not allowed routine
segments, we investigated the following experimental hypothesis H_{A1}: *Employing
SCAN, thus neglecting the manual identification stage of the not allowed real-word
routine behaviour through declarative constraints, is more effective than employing
traditional approaches (e.g. DISCO) that require to manually identify and filter*

---

[2] https://fluxicon.com/disco/

Table 5.1 Effectiveness of SCAN: p-values associated to each question

Q_{A1}		Q_{A2}		Q_{A3}	
DISCO	SCAN	DISCO	SCAN	DISCO	SCAN
5	4	5	4	4	5
4	4	5	3	4	5
4	3	5	3	3	5
4	2	5	3	3	5
4	2	4	2	2	5
3	2	4	2	2	4
3	2	4	1	1	4
2	2	3	1	1	4
1	2	2	1	1	4
p-value: 0.1443957		p-value: 0.0018155		p-value: 0.0005373	

out the not allowed routine segments. To validate H_{A1}, a *between-subject approach* was used, i.e., each user in p_{A1} (p_{A2}, respectively) was assigned to a different experimental condition, related to the exclusive use of SCAN (c_{A1}) or DISCO (c_{A2}) to perform the required steps for accomplishing both the case studies. Any user in p_{A1} was preliminarily instructed about the functionalities of SCAN throughout a short training session, while the users in p_{A2} already know how to use DISCO.

We evaluated the validity of H_{A1} by asking any user expert that completed the user study the following three questions:

- Q_{A1}: *The segment's filtering process required to filter out the not allowed routine segments is a complex task. Do you agree?*
- Q_{A2}: *The inspection of the routine segments is a complex task. Do you agree?*
- Q_{A3}: *SCAN (DISCO, respectively) makes the segmentation task feasible. Do you agree?*

Questions are rated with a 5-point Likert scale ranging from 1 ("strongly disagree") to 5 ("strongly agree"). To validate Q_{A1}, Q_{A2} and Q_{A3} we performed a comparison of the rates obtained from the questionnaire, respectively in the cases of c_{A1} and c_{A2}. Specifically, for each question, we employed a *2-Sample t-test* with a 95% confidence level to determine whether the means between the two distinct populations (i.e., p_{A1} and p_{A2}) involved in c_{A1} and c_{A2} differ. We measured the level of statistical significance by analyzing the resulting p-value. We remind that a $p-value \leq 0.05$ is considered to be statistically significant, while a $p-value \leq 0.01$ indicates that there is substantial evidence in favour of the experimental hypothesis. The results of the analysis are summarized in Table 5.1 that shows the values sorted in descending order, assigned to the responses of each user.

It appears evident that the experimental hypothesis H_{A1} is statistically supported by the results obtained for Q_{A2} and Q_{A3}, while it is rejected for Q_{A1}. Concerning Q_{A1}, it seems that the segment's filtering process was relatively easier in SCAN with respect to DISCO. Still, there is no statistical difference among the two distinct populations since for Q_{A1}, the p-value obtained is 0.1443957, which is greater than 0.05, and this means that hypothesis H_{A1} is rejected on Q_{A1}. On the other hand, the inspection of routine segments in DISCO seems to be more complex than SCAN since, for Q_{A2}, the p-value obtained is 0.0018155, which is less than 0.05, and this means that the hypothesis H_{A1} is accepted on Q_{A2}. Finally, for Q_{A3}, we got a p-value equal to 0.0005373, which is less than 0.05, and this means the hypothesis H_{A1} is accepted on Q_{A3}. In particular, this value is less than 0.01, meaning that there is a substantial difference between the means of the two distinct populations. This is reflected in higher values associated with SCAN and lower values associated with DISCO, thus making the segmentation task more feasible in SCAN with respect to DISCO. Therefore, H_{A1} can be considered partially accepted since it is validated for both Q_{A2} and Q_{A3} but rejected for Q_{A1}, where there is no statistical evidence that the use of SCAN is more effective than traditional process mining solutions (e.g., DISCO) in the process of segment's filtering.

5.3 Assessing the Robustness of SCAN

To investigate the robustness of SCAN regarding the achievement of user tasks specified in both Case Study #1 and Case Study #2, we collected the event logs generated during the user study. Subsequently, we compared these logs with the ground truth event logs, which were computed beforehand as results of the case studies. Specifically, the *robustness* is measured as the ratio between the number of logs compliant with the ground truth logs and the total number of logs, for both p_{A1} (i.e., SCAN) and p_{A2} (i.e., DISCO), grouped by Case (i.e., Case Study #1 and Case Study #2). In the following, we will show the results obtained both for Case Study #1 and for Case Study #2. Note that both the populations p_{A1} and p_{A2} first executed Case Study #1 in a limited time of 10 minutes and then Case Study #2, considered more complex, in 20 minutes.

- **Case Study #1**. Both p_{A1} and p_{A2} had 10 minutes to read the assigned track and perform the task either on DISCO (i.e., p_{A2}) or SCAN (i.e., p_{A1}) respectively. For p_{A2}, it is important to remember that users already know how to use the tool. The results obtained in this case show that 8 out of 9 people executed the task correctly, arriving at the correct event log, while 1 person obtained an incorrect result. Thus, the robustness in the case of p_{A2} is calculated as follows: $Robustness_{p_{A2}} = \frac{8}{9} = 0.88$.
  On the other hand, for p_{A1}, we remind the reader that the users experienced SCAN for the first time during this experimental session. In this case, the number of users who achieved the correct result is 6 out of 9, while 3 computed an incorrect event log. Therefore, the robustness in the case of p_{A1} is calculated as $Robustness_{p_{A1}} = \frac{6}{9} = 0.66$.
- **Case Study #2**. This case was executed immediately after the first one. The time allowed for achieving the task was 20 minutes due to the increased complexity compared to the previous one. For the group of users belonging to p_{A2}, the result obtained was that 4 out of 9 people computed the correct result while 5 computed the wrong one. It follows that the robustness in the case of p_{A2} is $Robustness_{p_{A2}} = \frac{4}{9} = 0.44$.

On the contrary, users assigned to p_{A1} performed much better. Indeed, 7 out of 9 users computed the correct result, while 2 computed the wrong one. As a consequence, the robustness for the users that used SCAN is $Robustness_{p_{A1}} = \frac{7}{9} = 0.77$.

If we compare the degree of robustness for both SCAN and DISCO in each case study, the following observations can be made:

- For Case Study #1, better results are achieved with DISCO. This is because the original log contains solely 8 routine variants, and among these only 4 were correct. For this reason, they were easily identifiable and therefore easy to be manually filtered. Regarding SCAN, we can say that since this was the first time the users experienced the tool, it is possible that the limited time of 10 minutes was not enough for completing the task. In addition, it is also possible that users had not yet settled into using SCAN even if they had been instructed during the short training session, thus before the user experiments.
- On the other hand, for Case Study #2, better results are achieved with SCAN. Since the original log presents more than 80 variants, manually identifying the wrong routine segments for filtering becomes even more challenging with DISCO. DISCO required users to filter the wrong routine segments one by one, whereas SCAN allows for the application of a limited number of declarative constraints to filter out a large number of wrong routine segments, thus bypassing the manual identification stage. Furthermore, the learning effect plays a crucial role in achieving good results, as users trained themselves while completing the task outlined in Case Study #1. This learning experience is reflected in the accomplishment of Case Study #2.

5.4 Quantifying the Usability of the UI of SCAN

We investigated the degree of *usability* of the UI developed for SCAN through the administration of the SUS (Software Usability Scale) questionnaire (which is one of the most widely used methodologies to measure the users' perception of the usability of a tool [90]) to the 9 PMM students that were involved in the experimental condition c_{A1}, i.e., that used SCAN. The questionnaire consists of 10 statements, adapted to SCAN and, evaluated with a Likert scale that ranges from 1 ("strongly disagree") to 5 ("strongly agree"):

- I think that I would like to use SCAN frequently.
- I found SCAN unnecessary complex.
- I thought SCAN was easy to use.
- I think that I would need the support of a technical person to be able to use SCAN.
- I found the various functions in SCAN well integrated.
- I thought there was too much inconsistency in SCAN.
- I would imagine that most people would learn to use SCAN very quickly.

Table 5.2 Computation of the SUS overall score.

User	q1	q2	q3	q4	q5	q6	q7	q8	q9	q10	SUS Score	Average
p1	5	1	5	1	5	1	5	1	5	1	100.0	82.5
p2	5	2	4	1	4	2	5	2	3	3	77.5	
p3	5	1	4	1	4	2	2	1	4	2	80.0	
p4	4	3	4	3	3	2	4	2	4	2	67.5	
p5	4	1	4	3	4	2	5	1	4	3	77.5	
p6	4	2	5	2	4	2	5	1	4	2	82.5	
p7	5	4	5	2	5	1	5	4	5	1	82.5	
p8	4	2	5	1	4	2	5	2	5	2	85.0	
p9	5	1	5	1	4	2	4	2	5	1	90.0	

- I found SCAN very awkward to use.
- I felt very confident using SCAN.
- I needed to learn a lot of things before I could get going with SCAN.

At the end of the questionnaire, an overall score is assigned to the questionnaire. To compute the SUS score for each PMM student, we need to determine each item's score contribution, which will range from 0 to 4. For odd items the score contribution is the scale position minus 1 $(x_i - 1)$. While for even items, the score contribution is 5 minus the scale position $(5 - x_i)$. To get the overall SUS score, multiply the sum of the item score contributions by 2.5. Thus, overall SUS scores range from 0 to 100 in 2.5-point increments. The score contributions can range from 0 to 40 (10 items with five scale steps ranging from 0 to 4). So, to obtain the multiplier necessary to increase the apparent range of the scale added to 100, divide 100 by the maximum sum of 40, equal to 2.5. Finally, to obtain the final SUS score, it is necessary to compute the average of those obtained by individual users. Table 5.2 represents all the values associated with the responses of the questions, the SUS score of each user, and the final SUS score.

The final SUS score can be compared with several benchmarks presented in the research literature to determine the degree of usability of the tool being evaluated. In our test, we made use of the benchmark given in [90], which associates to each range of the final SUS score a percentile ranking varying from 0 to 100, indicating how well it compares to other 5,000 SUS observations performed in the literature. The collection of the ranks associated with any statement of the SUS is reported in Table 5.2, calculated following the steps discussed in [90]. Since the final SUS score obtained by the tool was 82.5, according to the selected benchmark (see Fig. 5.2 taken from [90]), the usability of the tool corresponds to a rank of A, which indicates a degree of usability almost excellent.

The result shows that the UI implemented has been comprehensive and straightforward since the first use of the tool. And also that the use of the tool has been found effective and performing in achieving the required tasks.

Table 8.6 Curved Grading Scale Interpretation of SUS Scores

SUS Score Range	Grade	Percentile Range
84.1–100	A+	96–100
80.8–84	A	90–95
78.9–80.7	A–	85–89
77.2–78.8	B+	80–84
74.1–77.1	B	70–79
72.6–74	B–	65–69
71.1–72.5	C+	60–64
65–71	C	41–59
62.7–64.9	C–	35–40
51.7–62.6	D	15–34
0–51.7	F	0–14

Fig. 5.2 Curved Grading Scale Interpretation of final SUS Scores

Chapter 6
Routine Traces Detection through Trace Alignment

In this chapter, we present the last component of our approach to the automated segmentation of UI logs which exploits trace alignment in Process Mining to detect from a UI log all those user actions belonging to a valid routine segment (i.e., as output of the human-in-the-loop interaction step) and cluster them into well-bounded routine traces, thus achieving the segmentation task (**C1**).

The chapter is organized as follows. In Section 6.1, we provide an overview of the general approach to routine traces detection, depicting its main steps. Then, in Section 6.2, we delve into the technical details of the algorithm that implements the approach using the RPA use case discussed in Section 2.6.

6.1 The General Approach and the Routine Traces Detection Algorithm

The general approach to the routine traces detection consists of two methodological phases, *filtering* and *trace alignment*, to be applied in sequence, as shown in Fig. 6.1. Algorithm 1 shows the technical details of the algorithm that concretely implements such phases[1].

The algorithm takes in input a UI log U, a set of interaction models W_{set} and returns a set of routine-based logs U_{set}. For each interaction model $w \in W_{set}$ (one for each valid routine segment of interest) represented as Petri nets, the algorithm performs the following steps:

1. **Filtering**: The filtering phase is used to filter out noisy actions from the UI log. Specifically, for each interaction model $w \in W_{set}$, a local copy of the UI log U^w is created (line 3). Then, all user actions that appear in U^w but that can not be replayed by any transition of w are removed from U^w. The output of this step is a *model-based filtered UI log* U_ϕ^w (line 4). Working with U_ϕ^w rather than

---

[1] https://github.com/bpm-diag/SupSeg

S. Agostinelli: *Generating Executable Robotic Process Automation Scripts from Unsegmented User Interface Logs*, LNBIP 522, pp. 59–64, 2024.
https://doi.org/10.1007/978-3-031-61368-5_6

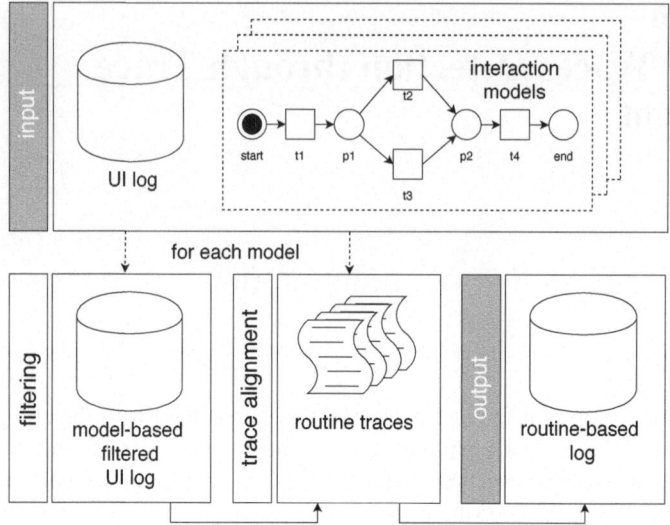

Fig. 6.1 Overview of the general approach underlying the routine traces detection component

with U^w will allow us to apply the trace alignment technique neglecting all the potential moves in log with user actions that could never be replayed by w. As a consequence, this will drastically reduce the number of alignment steps required to find optimal alignments, and at the same time optimize the performance of the algorithm. Before moving to the next step, a new routine-based log U_R^w is initialized (line 5).

2. **Trace Alignment**: The second step consists of applying the trace alignment technique discussed in Section 2.9 for any interaction model $w \in W_{set}$ and its associated model-based filtered UI log U_ϕ^w. This enables to extract from U_ϕ^w all those user actions that match a distinguishable pattern with w in the form of an optimal alignment γ^{opt} (line 7). Trace alignment allows to pinpoint the *synchronous moves* between U_ϕ^w and w. If they exist, the user actions involved in synchronous moves are extracted and stored into γ_{sm}^{opt} (line 8). Note that focusing just on synchronous moves allows us to exclude all redundant user actions from the analysis. Then, the algorithm:

a. creates a trace τ_{sm} consisting of the user actions associated with the synchronous moves stored in γ_{sm}^{opt} (line 10);
b. creates a (temporary) UI log U_{sm}^w containing only the trace τ_{sm} (line 11), which is required to properly run (again) trace alignment;
c. performs a new alignment between U_{sm}^w and w with the goal to compute the fitness value (line 12).

In case the fitness value is equal to 1, this means that the U_{sm}^w (and, consequently, τ_{sm}) can be replayed from the start to the final marking of w, making τ_{sm} a valid

---

Algorithm 1: Algorithm implementing the routine traces detection component

---

Parameters: a UI log U, a set of interaction models W_{set}
Result: A set U_{set} of routine-based logs

1   $U_{set} \leftarrow \emptyset$;
2   **forall** $w \in W_{set}$ **do**
3      $U^w \leftarrow$ duplicate(U);
4      $U^w_\phi \leftarrow$ filter(U^w);
5      $U^w_R \leftarrow \emptyset$;
6   **repeat**
7         $\gamma^{opt} \leftarrow$ *trace alignment* (U^w_ϕ, w);
8         $\gamma^{opt}_{sm} \leftarrow$ extract(γ^{opt});
9   **if** γ^{opt}_{sm} *is not empty* **then**
10   create a trace τ_{sm} from γ^{opt}_{sm};
11   create a temporary UI log U^w_{sm} from τ_{sm};
12            $fitness \leftarrow$ compute fitness from *trace alignment* (U^w_{sm}, w);
13   **if** $fitness$ *is 1* **then**
14   add τ_{sm} to U^w_R;
15   **else**
16   discard τ_{sm};
17   **end**
18   remove the events associated to τ_{sm} from U^w_ϕ;
19   **end**
20   **until** γ^{opt}_{sm} *is not empty*;
21   add U^w_R to U_{set};
22   **end**
23   **return** U_{set}

---

routine trace of w. In such a case, τ_{sm} is stored into U^w_R (line 14) and all the events associated to the synchronous moves in τ_{sm} are removed by U^w_ϕ (line 18). On the contrary, a fitness value lower than 1 indicates the presence of at least one move in the model in τ_{sm} with respect to w, i.e., τ_{sm} can not be completely replayed by w and is not a valid routine trace, meaning that we can discard it (line 16).

The above two steps can be repeated until γ^{opt}_{sm} is not empty (line 20), i.e., until there are synchronous moves in the computed alignment. At the end of the iteration, the routine-based log U^w_R is stored into U_{set} (line 21), and the algorithm starts to analyze the next interaction model into W_{set}. In conclusion, the algorithm computes a number of routine-based logs equal to the number of interaction models under study (associated to the valid routine segments).

It is worth to notice that: *(i)* for the computation of the trace alignment, the algorithm relies on the highly-scalable planning-based alignment technique implemented in [30]; and *(ii)* the routine traces detection component that exploits *trace alignment* in Process Mining can also be employed as a stand-alone supervised segmentation technique as described in [11], under the assumption to know a priori the interaction models of the routines to identify in the UI log (cf. [71]). In this setting, the technique is able to achieve all variants of cases 1, 2, and (partially) 3, except when there are interleaved executions of shared user actions of many routines. In that case, the risk exists that a shared user action is associated with a wrong routine execution (i.e., Case 3.3 and Case 3.4 are not covered). While in [11], to make the technique works, it is required to know at the outset the structure (i.e., the interaction models) of the routines to identify in the UI log, in [3] we have mitigated this assumption by semi-automatically discovering such structures in the form of routine segments, and then used them as input for the routine traces detection component, since the approach is not aware of the concrete behaviour of the routines of interest, which the approach itself will discover, but instead exploits the end-delimiters associated to any user action that ends a routine execution.

6.2 An Execution Instance of the Routine Traces Detection Algorithm

We show now an execution instance of Algorithm 1 applied to the original UI log generated from the case study of Section 2.6:

$U = \{A, B, C_{11}, D_{11}, E_{11}, F_{11}, G_{11}, H_{11}, I_{11}, L_{11}, B, M_{21}, N_{21}, B, C_{12}, D_{12}, E_{12}, F_{12}, G_{12}, H_{12}, I_{12}, L_{12}, B, M_{22}, O_{22}, \ldots, A, B, C_{1(i-1)}, Y_1, D_{1(i-1)}, E_{1(i-1)}, F_{1(i-1)}, G_{1(i-1)}, G_{1(i-1)}, G_{1(i-1)}, H_{1(i-1)}, I_{1(i-1)}, L_{1(i-1)}, B, M_{2(i-1)}, N_{2(i-1)}, B, Y_{n-1}, C_{1i}, D_{1i}, E_{1i}, Y_n, F_{1i}, G_{1i}, H_{1i}, I_{1i}, I_{1i}, I_{1i}, L_{1i}, B, M_{2i}, O_{2i}\}$.

The log contains elements of noise, i.e., user actions $Y_{k \in \{1,n\}}$ that are not allowed by routine segments W_1, W_2, W_3 and W_4, and redundant actions like G and I that are unnecessary repeated multiple times. In addition, A and B are shared user actions, as they are included in the interaction models of both R_1 and R_2. In particular, A is potentially involved in the enactment of any execution of R_1 and R_2, while B is required by all executions of R_1 and R_2.

The algorithm takes in input: *(i)* the UI log U and *(ii)* the interaction models of W_1, W_2, W_3 and W_4, and computes a set of routine-based logs U_{set} by executing the following steps:

- (line 1): initializes the set of interaction models U_{set};
- (line 2): iterates on the interaction models of W_1, W_2, W_3 and W_4. For the sake of simplicity, we focus only on the steps computed in the case of W_1;
- (line 3): creates a local copy of U, namely U^w;
- (line 4): filters U^w from noise, so $U_\phi^w = \{A, B, C_{11}, D_{11}, E_{11}, F_{11}, G_{11}, H_{11}, I_{11}, L_{11}, B, B, C_{12}, D_{12}, E_{12}, F_{12}, G_{12}, H_{12}, I_{12}, L_{12}, B, \ldots, A, B, C_{1(i-1)}, D_{1(i-1)}, E_{1(i-1)}, F_{1(i-1)}, G_{1(i-1)}, G_{1(i-1)}, G_{1(i-1)}, H_{1(i-1)}, I_{1(i-1)}, L_{1(i-1)}, B, B, C_{1i}, D_{1i}, E_{1i}, F_{1i}, G_{1i}, H_{1i}, I_{1i}, I_{1i}, I_{1i}, L_{1i}, B\}$.
  In this step, the user actions $Y_{k \in \{1,n\}}$ and M, N, O (being exclusively related to W_3 and W_4) are filtered out by the log. On the other hand, redundant actions still remain in the log;
- (line 5): initializes the routine-based log U_R^w;

- (line 7): computes the trace alignment between U_ϕ^w and the interaction model of W_1, namely w.

$$\begin{array}{|c|c|c|c|c|c|c|c|c|c|c|c|}\hline A & B & C_{11} & D_{11} & E_{11} & F_{11} & G_{11} & H_{11} & I_{11} & L_{11} & B & ... \\ \hline A & B & C & D & E & F & G & H & I & L & \gg & ... \\ \hline \end{array}$$

- (line 8): extracts the synchronous moves from γ^{opt} into γ_{sm}^{opt}.
- (line 9): evaluates to $True$, as γ_{sm}^{opt} is not empty;
- (line 10): computes the trace τ_{sm} starting from γ_{sm}^{opt}. So $\tau_{sm} = \langle A, B, C_{11}, D_{11}, E_{11}, F_{11}, G_{11}, H_{11}, I_{11}, L_{11}\rangle$;
- (line 11): adds the trace τ_{sm} in U_{sm}^w;
- (line 12): computes trace alignment between U_{sm}^w and w.

$$\begin{array}{|c|c|c|c|c|c|c|c|c|c|}\hline A & B & C_{11} & D_{11} & E_{11} & F_{11} & G_{11} & H_{11} & I_{11} & L_{11} \\ \hline A & B & C & D & E & F & G & H & I & L \\ \hline \end{array}$$

U_{sm}^w can be replayed without deviations from the start to the final marking of w, meaning a perfect fitness between the log and the interaction model;

- (line 13): evaluates to $True$, as the fitness of the alignment (cf. line 12) is equal to 1;
- (line 14): adds τ_{sm} in U_R^w, i.e., τ_{sm} is recognized as a valid routine trace;
- (line 18): removes all the events associated with the synchronous moves in τ_{sm} from U_ϕ^w. Thus, $U_\phi^w = \{B, B, C_{12}, D_{12}, E_{12}, F_{12}, G_{12}, H_{12}, I_{12}, L_{12}, B, \ldots, A, B, C_{1(i-1)}, D_{1(i-1)}, E_{1(i-1)}, F_{1(i-1)}, G_{1(i-1)}, G_{1(i-1)}, G_{1(i-1)}, H_{1(i-1)}, I_{1(i-1)}, L_{1(i-1)}, B, B, C_{1i}, D_{1i}, E_{1i}, F_{1i}, G_{1i}, H_{1i}, I_{1i}, I_{1i}, I_{1i}, L_{1i}, B\}$;
- (line 20): Since γ_{sm}^{opt} is not empty, the algorithm comes back to line 6. After repeating the above steps from line 7 to line 14, the algorithm computes the following alignment:

$$\begin{array}{|c|c|c|c|c|c|c|c|c|c|c|c|c|}\hline ... & B & A & B & C_{1(i-1)} & D_{1(i-1)} & E_{1(i-1)} & F_{1(i-1)} & G_{1(i-1)} & G_{1(i-1)} & G_{1(i-1)} & H_{1(i-1)} & I_{1(i-1)} & L_{1(i-1)} & B & ... \\ \hline ... & \gg & A & B & C & D & E & F & G & \gg & \gg & H & I & L & \gg & ... \\ \hline \end{array}$$

and discovers a second routine trace $\tau_{sm} = \langle A, B, C_{1(i-1)}, D_{1(i-1)}, E_{1(i-1)}, F_{1(i-1)}, G_{1(i-1)}, H_{1(i-1)}, I_{1(i-1)}, L_{1(i-1)}\rangle$ and adds it in U_R^w. Like before, all the events associated with the synchronous moves in τ_{sm} are removed from U_ϕ^w. Thus, $U_\phi^w = \{B, B, C_{12}, D_{12}, E_{12}, F_{12}, G_{12}, H_{12}, I_{12}, L_{12}, B, \ldots, G_{1(i-1)}, G_{1(i-1)}, B, B, C_{1i}, D_{1i}, E_{1i}, F_{1i}, G_{1i}, H_{1i}, I_{1i}, I_{1i}, I_{1i}, L_{1i}, B\}$.

The subsequents iterations of the algorithm do not discover new routine traces for W_1. In particular, the alignment steps between w and U_ϕ^w are all moves in the log since all the As are already extracted. It is worth to notice that redundant user actions G and I are removed from U_ϕ^w during these iterations. The algorithm ends to iterate when γ_{sm}^{opt} is empty, that is, when there are no more synchronous moves to extract;

- (line 21): After the last iteration ends, the routine-based log U_R^w is stored into U_{set}, and the algorithm starts to analyze the interaction model of W_2.

The outcome of the segmentation task will be a set of routine-based logs U_{set} (in this case four, since the number of interaction models under study is four) generated as follows:

- $U_{W_1} = \{\langle A_{11}, B_{11}, C_{11}, D_{11}, E_{11}, F_{11}, G_{11}, H_{11}, I_{11}, L_{11} \rangle, \ldots, \langle A_{1(i-1)}, B_{1(i-1)}, C_{1(i-1)}, D_{1(i-1)}, E_{1(i-1)}, F_{1(i-1)}, G_{1(i-1)}, H_{1(i-1)}, I_{1(i-1)}, L_{1(i-1)}, \rangle\}$
- $U_{W_2} = \{\langle B_{12}, C_{12}, D_{12}, E_{12}, F_{12}, G_{12}, H_{12}, I_{12}, L_{12}, \rangle, \ldots, \langle B_{1i}, C_{1i}, D_{1i}, E_{1i}, F_{1i}, G_{1i}, H_{1i}, I_{1i}, L_{1i} \rangle\}$
- $U_{W_3} = \{\langle B_{21}, M_{21}, N_{21} \rangle, \ldots, \langle B_{2(i-1)}, M_{2(i-1)}, N_{2(i-1)} \rangle\}$
- $U_{W_4} = \{\langle B_{22}, M_{22}, O_{22} \rangle, \ldots, \langle B_{2i}, M_{2i}, O_{2i} \rangle\}$

Commercial RPA tools can employ routine-based logs to synthesize executable scripts in the form of SW robots without the manual modeling of the routines. To this end, the SmartRPA approach presented in the next chapter is able to automatically synthesize executable scripts for enacting SW robots at run-time directly from the routine-based logs.

Part III
Automated Generation of SW Robots

Chapter 7
Defining and Designing SmartRPA

RPA solutions access the UI layer of SW applications and provide a virtual workforce of SW robots that can mimic human keyboard and mouse interactions with a UI as if a real person was doing them. To take full advantage of this technology, organizations leverage the support of skilled human experts that preliminarily observe how routines are executed on the UI of the involved SW applications and then implement the executable RPA scripts required to automate the routines enactment by SW robots on a target computer system. However, the current practice is time-consuming and error-prone, as it strongly relies on the ability of the human experts to correctly interpret the routines (and their variants) to automate.

Although RPA is generally considered an easy to implement technology, in-depth knowledge is necessary to create reliable and scalable SW robots, particularly when intermediate user inputs are required to progress the execution of a routine properly. As a result, between 30% and 50% of initial RPA implementations are estimated to fail [86, 58]. Consequently, an approach that simplifies the realization of an RPA project, particularly the generation of SW robots in the presence of many routine variants, can be considered a relevant artefact to investigate. This leads to the following research questions:

- **RQ2.1**: Which steps are required to make the generation of SW robots less dependent by the intervention of RPA human experts?
- **RQ2.2**: How can the detection of variants (and related variation points) in a routine be automatically achieved?
- **RQ2.3**: What is the effectiveness of employing an approach that synthesizes SW robots neglecting the (manual) specification stage of the routines' behaviour through flowchart models?

In answering these questions, we contribute to three recent challenges related to **C2** that were put forward in [9, 10, 25, 63, 6], namely: **(C2.1)** the automated identification of the routine steps to robotize from a UI log, **(C2.2)** the automated detection of all the routine variants that require some user input to proceed with their execution, and **(C2.3)** the automated synthesis of executable RPA scripts for enacting SW robots at run-time. The result is an approach and an implemented tool, called

© The Author(s), under exclusive license to Springer Nature Switzerland AG 2024
S. Agostinelli: *Generating Executable Robotic Process Automation Scripts from Unsegmented User Interface Logs*, LNBIP 522, pp. 67–80, 2024.
https://doi.org/10.1007/978-3-031-61368-5_7

SmartRPA, which is able to *(i)* interpret the UI logs recording the mouse/key events that happen on the UI of the SW applications involved in many routine executions, *(ii)* discover all the variants (and variation points) of the routine under observation, and *(iii)* automatically combine them into an executable RPA script, which can be *reactively* synthesized into a single SW robot.

Differently from the literature approaches to automated RPA scripts generation from UI logs (cf. Section 7.2), which enable to automate straightforward routines that have essentially no variance and do not require any human intervention, the SW robots generated by SmartRPA are obtained to handle the intermediate user inputs that are required during the routine execution, thus enabling to emulate the most suitable routine variant for any specific combination of user inputs as observed in the UI log. This makes the synthesis of SW robots performed by SmartRPA *reactive* to any user decision found during a routine execution. "Reactivity" highlights the fact that the behaviour of SW robots is determined immediately before their enactment, as it is driven by the specific user inputs required to execute the routine. This also means that reactivity enable the potential run-time generation of as many SW robots as are the different variants of the routine to be emulated. Therefore, SmartRPA acknowledges the benefit of human involvement at multiple points of the routine execution, leveraging the "human-in-the-loop" model for the automated execution of routines that are less static and require variable decisioning [25].

We structure this Chapter according to the activities suggested by Johannesson and Perjons in [52] for delivering a design science artefact. Specifically, Section 7.1 describes our research methodology. Section 7.2 discusses the related work solutions to the research challenges, with the aim to derive the technical requirements for the design of the SmartRPA approach, whose main steps are examined in Section 7.3. Section 7.4 outlines the algorithm for the automated detection of variation points from many routine executions.

7.1 Research Methodology

Our research methodology is inspired to the Design Science approach described by Johannesson and Perjons in [52]. The methodology is applied in four distinct sequential phases: problem formulation and objectives, requirements definition, design and development, and demonstration and evaluation. See Fig. 7.1 for an overview.

Problem Formulation and Objectives. In this phase, which is already tackled at the beginning of the Chapter, we first identify and specify the research problem to be tackled, i.e., the *reactive synthesis of SW robots in an automated way from UI logs*. Then, we justify its significance in the RPA field. The relevance of the problem is also supported by the presence of three related *research challenges*, i.e., **C2.1**, **C2.2** and **C2.3**, taken from previous works [9, 10, 25, 63, 6]. Finally, we elaborate three main *research questions*, i.e., **RQ2.1**, **RQ2.2** and **RQ2.3**, for guiding our research towards the definition of an *artefact* to solve the problem. Such an artefact is represented by an approach and an implemented tool, called SmartRPA, which is able to interpret the

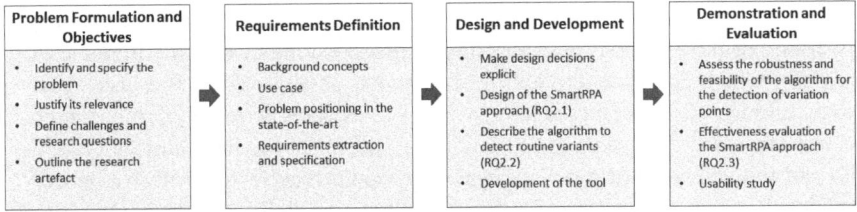

Fig. 7.1 Research methodology based on Johannesson and Perjons [52]

UI logs keeping track of many routine executions, and automatically and reactively synthesize SW robots that emulate the most suitable routine variant for any specific intermediate user input that is required during the routine execution.

Requirements Definition. The second phase consists of eliciting the requirements for the outlined artifact. After providing the necessary background concepts in Chapter 2 on routines, SW robots, and UI logs, along with a real-life RPA use case, we investigate the related work in Section 7.2. This includes examining documented solutions to similar research challenges, in order to extract the technical requirements needed to support the design and development of SmartRPA.

Design and Development. Based on the analysis of the related work and the derived technical requirements, in the third phase, we make design decisions explicit, discussing the SmartRPA approach and describing its stages to answer **RQ2.1**. Moreover, we present in detail a novel algorithm to automatically identify routine variants and variation points from UI logs, thus addressing **RQ2.2** and enabling a reactive synthesis of the SW robots. Lastly, we show the technical steps enacted to develop the SmartRPA approach as a real implemented tool.

Demonstration and Evaluation. In the fourth phase, to understand the general quality of SmartRPA to tackle the research challenges, we analyze four *non-functional requirements* on the artefact. Specifically, we first perform many synthetic experiments employing UI logs of increasing complexity to assess the *robustness* and *feasibility* of our approach to the identification of routine variants and variation points for the reactive synthesis of SW robots. Then, to answer **RQ2.3**, we perform a controlled experiment involving real users exploiting the RPA use case of Section 2.6 to investigate the *effectiveness* of the SmartRPA approach when compared to a traditional model-based approach for the generation of SW robots. Finally, we quantify the *usability* of the UI provided by the tool implementing the SmartRPA approach.

7.2 Related Work Analysis and Requirements Specification

Most commercial RPA tools enable RPA user experts to tag the variation points directly in the flowchart model of the routine under study. That is, it can be modeled and properly emulated by a SW robot if an in-depth knowledge of the anatomy and working of the routine is available during the modeling task. But without such knowledge, which is based on careful observation sessions of human users that perform routine tasks in their computer systems, it becomes extremely complex both to identify the candidate steps of the routine to specify in the flowchart model (cf. **C2.1**) and the detection of those variants that would require some user inputs to proceed with their execution (cf. **C2.2**). In a nutshell, the ability of commercial RPA tools to emulate all the possible behaviours of the routine depends on the correctness of the modeling task, without which it is not possible to automatically generate the executable RPA scripts to be embedded into the SW robots (cf. **C2.3**).

In this direction, this section presents the relevant approaches from the research literature that are able to mitigate the above challenges by skipping the modeling activity of the flowchart diagram. Then, in an attempt to fully address them, we derive a set of technical requirements from realizing our SmartRPA approach.

Specifically, the research literature proposes many approaches that are targeted to automatically discover and implement the behaviour of SW robots by interpreting the working of the routines stored into previously recorded UI logs. Towards the addressing of **C2.1**, the works [64, 22] provide an approach, coupled with an implemented tool, that leverages process mining techniques to *(i)* keep track of UI actions performed within Excel and Google Chrome into an event log, and *(ii)* extract the fragments of a routine that can be eventually automated by a third-party RPA tool. Similarly, in [67] it is presented the *Desktop Activity Mining* tool, which is able to record the user actions performed during an office-based routine task on a UI and to discover a process model describing the behaviour of such routine. Note that the proposed tool is based on recording the mouse click coordinates on the screen and storing them in a dedicated UI log. Thus it can not replicate the same user's observed behaviour performed in different computer systems, lacking portability.

Even if the works [64, 22, 67] do not tackle the issue of synthesizing executable RPA scripts from the identified candidate routines, they had the vision that the behaviour of a routine can be inferred by observing and interpreting the footprint of the routine itself from a UI log that keeps track of its user actions. This directly leads to three technical requirements that need to be met to tackle **C2.1**:

Req₁ - Recording UI Logs: A feature to record the low-level user actions executed during one (or many) routine(s) enactment on the UI in the form of a UI log is strongly needed to keep track of its behaviour.

Req₂ - Extraction of Routine-based Logs from a UI Log: A UI log may contain interleaved executions of one/many routine/s. As the target is to reason on the behaviour of a single routine per time, it is needed to pre-process the UI log to: *(i)* identify which user actions contribute to which routines inside the UI log; *(ii)* organize such actions into well-bounded routine traces and *(iii)* store them into a dedicated routine-based log.

Req₃ - Events Abstraction: A routine-based log is characterized by low-level user actions, and thus may contain noise and redundant actions that must be filtered out from the log itself.

With the aim to tackle **C2.2**, in [40], the authors propose a self-learning approach to detect high-level RPA-rules from historical low-level behaviour logs automatically. An if-then-else deduction logic is used to infer rules from behaviour logs by learning relations between the different routines performed in the past. Then, such rules are employed to facilitate the SW robots' instantiation. A similar approach is adopted in [59], where the *FlashExtract* framework is presented. FlashExtract allows for the extraction of relevant data from semi-structured documents using input-output examples, from which one can derive the relations underlying the working of a routine. Finally, in [79] the authors identify repetitive edits to text documents by keeping track of a graph of edits and suggest automation rules for SW robots.

The above works have provided a relevant contribution for the semi-automatically detection of variation points of a routine, with the aim to support the manual development of SW code by RPA expert users. In the direction of realizing a fully automated approach to the detection of routine variants and variation points, we can derive the following requirement:

Req$_4$- **Automated Detection of Variation Points**: An algorithm that is able to automatically detect the variation points of a routine from a routine-based log is required to reactively generate SW robots that correctly emulate the routine's behaviour.

Concerning **C2.3**, the literature proposes only a relevant solution, called Robidium [61], that tackles this challenge. Robidium is an approach and an open-source tool that enables to generate executable scripts (by only interpreting UI logs) that the commercial RPA tool UI Path[1] can enact.

The main feature of Robidium is that it automates only the most frequent routine variant among the ones discovered in the UI log. This is because Robidium synthesizes RPA scripts that do not require intermediate user inputs during their execution, i.e., it is focused on the generation of unattended SW robots. On the other hand, to synthesize attended SW robots, the following requirement is needed:

Req$_5$- **Automated and Reactive Generation of SW Robots**: A solution that is able to automatically and reactively synthesize RPA scripts is required for the generation of attended SW robots able to enact the most suitable routine variant depending on the specific input conditions at hand.

It is worth noticing that another group of approaches exists towards SW robots automation, which focuses on learning the structure of a routine from natural language descriptions of the procedure underlying the routine itself. In this direction, the work [49] defines a new grammar for complex workflows with chaining machine-executable meaning representations for semantic parsing. In [65], the authors provide an approach to learn activities from text documents employing supervised machine learning techniques such as feature extraction and support vector machine training. Similarly, in [44] the authors adopt a deep learning approach based on Long Short-Term Memory (LSTM) recurrent neural networks to learn the relationship between

---

[1] www.uipath.com

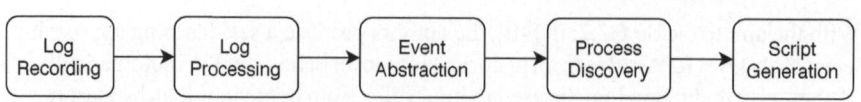

Fig. 7.2 Overview of the SmartRPA approach

activities of a routine task. The above works assume the availability of textual doc-
umentation of suitable quality and completeness at the outset, and neglect the fact
that users can perform steps in a routine that are not fully documented to deal with
variations and exceptions. This may potentially lead to imprecise results in describing the routine's anatomy. Therefore, these works seem to be particularly suitable
for discovering the desired structure of a routine, in contrast with the observed one,
like happens in all the log-based approaches discussed so far.

Finally, a third group of approaches exist that aim to eliminate human-dependent
training [16, 50]. They rely on probabilistic and machine learning algorithms to
automatically train SW robots to avoid any manual effort. These approaches are
currently the least mature if compared with the others discussed above, but potentially
with the best promises for realizing fully, automated intelligent RPA approaches.

7.3 Design of the SmartRPA Approach

From a methodological perspective, SmartRPA has been conceptualized and designed towards addressing the five technical requirements discussed in Section 7.2.
In addition, the approach underlying SmartRPA takes inspiration from the RPM
(Robotic Process Mining) framework presented by Leno et al. in [63]. RPM aims
to support analysts in producing executable specifications of routines in the form
of SW robots, interpreting the routine executions stored in a UI log. Specifically,
RPM envisions a pipeline of three main stages that consist of: *(i)* collecting and pre-
processing UI logs corresponding to executions of one or more routine executions;
(ii) identifying and discovering candidate routines to be automated with RPA tools;
and *(iii)* synthesizing executable RPA scripts. Robidium [61] is a concrete example
of how to realize the RPM approach.

To address the technical requirements, SmartRPA incorporates the three main
stages of the RPM framework within a larger approach that includes five operational
steps to be applied in sequence: *(i)* Log Recording, *(ii)* Log Processing, *(iii)* Event
Abstraction, *(iv)* Process Discovery, and *(v)* Script Generation, as shown in Fig. 7.2.
Note that such methodological steps are useful not only to tackle the technical
requirements, but also serve as our answer to **RQ2.1**.

Log Recording. SmartRPA belongs to those approaches that learn how to automate
routines "by examples". Therefore, a UI log that keeps track of the low-level user
actions generated during the interaction with the UIs of multiple SW applications
within the execution of a routine is required to derive its structure. To this end, a

training session in which several users perform the same routine to be automated is necessary to record the specific UI actions involved in its execution correctly. While there exist many monitoring and recording solution in the Human-Computer Interaction field [33] that keep track of the actions that a user is doing on the screen of a computer system (recorded as mouse click coordinates) during a controlled experiment, in SmartRPA we need dedicated recording features to produce a raw UI log corresponding to many executions of the same routine during a pre-defined period of time (cf. **Req₁**). At the end of the training session, the outcome of this step will consist of as many UI logs as the users who performed the routine under analysis from the start to the end.

Log Processing. It comes into play to pre-process the recorded UI logs and make them suitable for being correctly interpreted. Since any UI log obtained from the previous step keeps track of single, independent execution of the observed routine, a merging activity is needed to combine them into a single, larger UI log. In a nutshell, the content of any recorded UI log obtained after a training session will be considered as a single trace of the (larger) UI log being generated. Thus, if compared with the description of UI logs performed in Section 7.1, we can say that SmartRPA enables to interpret UI logs that are *routine-based*, i.e., logs that can be already considered as *well segmented* since the enactment of any training session will be represented by a specific routine trace in the log (cf. **Req₂**). Of course, this does not prevent noise and redundant user actions in the recorded routine traces, whose presence will be reduced in the next steps.

Event Abstraction. This step is targeted to convert the routine-based UI log (that will be later employed to generate the executable RPA scripts), which contains the low-level user actions recorded during the interaction with the UI, into a high-level version. Such a high-level version can be used for diagnostic and analysis purposes by expert RPA analysts to *(i)* filter out noise, i.e., irrelevant events for the routine execution. For instance, applications related to the operating system such as *Windows Updater* may start automatically while the UI log is being recorded, and they may dirty the recording phase of the users during their training session, thus they need to be filtered out; *(ii)* group similar events, mitigating noise (cf. **Req₃**).

Moreover, the high-level routine-based UI log will be used to derive the flowchart representing the abstract workflow describing the routine behaviour, employing dedicated high-level descriptive labels to define the high-level activities. We notice that the Event Abstraction step is an addition with respect to the RPM framework, which instead focuses on discovering the anatomy of a routine only for execution purposes.

Process Discovery. This step has a twofold objective:

- It takes in input the high-level routine-based log generated by the Event Abstraction component to derive the workflow describing the users' observed behaviour in the UI. An RPA analyst can analyze this workflow to look at the high-level structure of the routine under analysis.
- Moreover, the knowledge of the workflow underlying the routine, coupled with the low-level version of the routine-based UI log, will be used to detect the most suitable routine variant according to the intermediate user inputs observed into

the UI log and its encoding into a SW robot. In Section 7.4, it is reported a detailed discussion about the algorithm implemented to the identification of the routine variants and the related variation points (cf. **Req**$_4$), necessary to obtain a reactive synthesis of the SW robot.

It is worth noticing that in the RPM framework, the Process Discovery step shown above is (in part) realized through the second stage of the framework, targeted to the identification and discovering of the candidate routines to be automated. In SmartRPA, the knowledge about which routine has to be automated is already embedded into the UI logs obtained by performing the training session at the outset. Moreover, during the Event Abstraction step, the routine is further cleaned from noise and redundancies, keeping just the events in the UI that can contribute to the routine automation.

Script Generation. First of all, this step allows an RPA analyst to personalize the values stored in the events of the most suitable routine variant detected in the previous step before the generation and enactment of the SW robot. Finally, taking into account the edits made, SmartRPA can generate the required executable RPA script to run the SW robot that emulates the most suitable routine execution on the UI, by scanning the recorded low-level events stored in the routine-based UI log and converting them into executable pieces of SW code (cf. **Req**$_5$).

7.4 Automated Detection of Variation Points of a Routine

To properly address **RQ2.2**, in this section, we present an algorithm to identify different variation points of a routine by inspecting multiple executions of the routine itself inside the low-level routine-based log obtained as the outcome of the Event Abstraction step. We remind the reader that *variation point* is a point in the routine execution where a user choice needs to be made between multiple possible variants (cf. Section 2.7). Identifying variation points is fundamental to synthesize SW robots that emulate the most suitable routine variants in relation to the intermediate user inputs provided during the routine enactment. To be more specific, Algorithm 2 takes in input the low-level routine-based log and builds in output a new routine-based log that categorizes the user actions that contribute to the identification of a new routine variant, distinguishing them from the (other) actions that are common to any routine trace recorded in the UI log. In the following, we discuss the main steps of Algorithm 2 relying on the RPA use case explained in Section 2.6. In particular, we will refer with $R_{example}$ the interaction model of the routine procedure obtained from connecting with an invisible transition the place "end" of R_1 to the place "start" of R_2 (cf. Fig. 2.4 and Fig. 2.5).

Algorithm 2: Reactive Synthesis of SW Robots

Parameters: A DataFrame df
Result: A DataFrame resultDF

```
 1  df ← align rows;                                            // Aligning step
 2  df["duplicated"] ← mark duplicated rows;                    // Marking step
 3  resultDF = None;                                      // store final groups
 4  previousDecidedDF = None;
 5  groups ← group rows of df;                                  // Grouping step
 6  for groupDF in groups do // Iterating step
 7  │   if groupDF.duplicated == True then // No decision to take
 8  │   │   rows ← get the rows of the first trace from groupDF;
 9  │   │   resultDF.append(rows);
10  │   end
11  │   else // variation point
12  │   │   if previousDecidedDF then
13  │   │   │   IDs ← IDs of traces compliant with previousDecidedDF;
14  │   │   │   filteredDF ← rows with case ID in IDs from groupDF;
15  │   │   │   decisionDF ← remove redundant rows from filteredDF;
16  │   │   end
17  │   │   else
18  │   │   │   decisionDF ← remove redundant rows from groupDF;
19  │   │   end
20  │   │   decisionDialog ← show decision dialog built from decisionDF;
21  │   │   decidedDF ← rows of decisionDF selected in decisionDialog;
22  │   │   resultDF.append(decidedDF);          // append rows from decidedDF
23  │   │   previousDecidedDF ← decidedDF;           // save current decision
24  │   end
25  end
26  return resultDF
```

Aligning (line 1)

The first step of the algorithm consists of aligning the different executions recorded in the routine-based log to make them more similar from each other, when possible. This means, on the one hand, removing user actions that are irrelevant for the execution of the SW robot, such as special URLs like about:blank or chrome://newtab/ or low-level events such as enableBrowserExtension or afterCalculate. And, on the other hand, identifying and moving in the same point of any trace of the log those sequences of events underlying exactly the same behaviour in different traces (e.g., copy/paste activities from a specific cell to a specific text field) but originally located in different points among the various traces.

Marking (line 2)

It is crucial to identify which variation points must be considered within many routine executions to enact the most suitable routine variant. To this end, the very first step consists of marking as "duplicated" those rows that underly exactly the same event performed on the UI in different routine traces of the log. In a nutshell, the *it-h* row of a routine trace is considered as duplicated if it includes an event that is the same in the *it-h* row of all the other routine traces. We evaluate two events as identical if the following data fields have the same value for the event in the *it-h* row in all the recorded routine traces:

- *category*: represents the category of the user action, e.g., *Browser*, *OperatingSystem*, *Clipboard*, and *MicrosoftOffice*;
- *application*: name of the application where the user action occurred, e.g., Google Chrome, Microsoft Excel, etc.;
- *concept:name*: name of the user action recorded by the Log Recording component;
- *event_src_path*: source path in the operating system related to a user action. It could indicate the path of a file or folder opened, modified, created or deleted. It could also denote the path of an executable program that has been opened or closed;
- *event_dest_path*: destination path in the operating system related to a user action. If a file or folder is renamed, the new path name is present in this column;
- *browser_url_hostname*: hostname of the url recorded within a routine-based log. Two rows could have different URLs but the same hostname (e.g., *www.uniroma1.it/students* and *www.uniroma1.it/contacts* both have *uniroma1.it* as hostname);
- *xpath*: *XML Path Language* is a query language for selecting nodes in a page. It is used to uniquely identify a HTML element in a webpage.

Only the above subset of data fields associated with an event in the UI log is evaluated to detect duplicate rows, because some data fields are always different among the several executions of a routine. For example, the *case ID* of a routine trace or its *timestamp* is unique and would always lead to false results if they were considered because all the rows would always result different (i.e., not duplicated), leading to the identification of many wrong variation points.

To sum up, if an event appears in all the routine traces of a routine-based log and the rows associated with that event have the same values for all the data fields discussed above, it means the users consistently executed the same user action on the UI at a specific point of the routine execution during their training session. Consequently, such "duplicated" events would not lead to any variation point of the routine.

On the other hand, if there exists at least a user that performed an action on the UI in the *i-th* step of a routine trace that differs (according to the data fields listed above) from the actions performed at the same *i-th* step of the other routine executions, then it means that the associated event only appears in certain routine traces and not in others, thus identifying a variation point. As a consequence, we mark as "not duplicated" all the *i-th* rows of any routine trace under analysis. From a technical point of view, a new column accepting boolean values called *duplicated* is added to the routine-based log. Fig. 7.3 shows a fragment of the user actions belonging to 3 different routine traces of $R_{example}$ identifying a variation point that leads to three

	case:concept:name	category	application	concept:name	browser_url	tag_value	duplicated
0	1109155433758000	MicrosoftOffice	Microsoft Excel	openWorkbook			True
1	1109155433758000	Browser	Chrome	startDownload	about:blank		True
2	1109155433758000	Clipboard	Clipboard	copy			True
3	1109155433758000	MicrosoftOffice	Microsoft Excel	editCell			True
4	1109155433758000	Browser	Chrome	changeField	https://bpm-diag.github.io/form/	Vanessa Costa	True
5	1109155433758000	Browser	Chrome	clickRadioButton	https://bpm-diag.github.io/form/	car_no	False
6	1109155433758000	Browser	Chrome	clickButton	https://bpm-diag.github.io/form/		True
7	1109155433758000	Browser	Chrome	submit	https://bpm-diag.github.io/form/		True
8	1109155433758000	Browser	Chrome	formSubmit	https://bpm-diag.github.io/form/		True
9	1109155658364000	MicrosoftOffice	Microsoft Excel	openWorkbook			True
10	1109155658364000	Browser	Chrome	startDownload	about:blank		True
11	1109155658364000	MicrosoftOffice	Microsoft Excel	editCell			True
12	1109155658364000	Clipboard	Clipboard	copy			True
13	1109155658364000	Browser	Chrome	changeField	https://bpm-diag.github.io/form/	Vanessa Marino	True
14	1109155658364000	Browser	Chrome	clickRadioButton	https://bpm-diag.github.io/form/	car_yes	False
15	1109155658364000	Browser	Chrome	clickRadioButton	https://bpm-diag.github.io/form/	car_accept	False
16	1109155658364000	Browser	Chrome	clickButton	https://bpm-diag.github.io/form/		True
17	1109155658364000	Browser	Chrome	submit	https://bpm-diag.github.io/form/		True
18	1109155658364000	Browser	Chrome	formSubmit	https://bpm-diag.github.io/form/		True
19	1111145627144000	MicrosoftOffice	Microsoft Excel	openWorkbook			True
20	1111145627144000	Browser	Chrome	startDownload	about:blank		True
21	1111145627144000	MicrosoftOffice	Microsoft Excel	editCell			True
22	1111145627144000	Clipboard	Clipboard	copy			True
23	1111145627144000	Browser	Chrome	changeField	https://bpm-diag.github.io/form/	Paola Valentini	True
24	1111145627144000	Browser	Chrome	clickRadioButton	https://bpm-diag.github.io/form/	car_yes	False
25	1111145627144000	Browser	Chrome	clickRadioButton	https://bpm-diag.github.io/form/	car_reject	False
26	1111145627144000	Browser	Chrome	clickButton	https://bpm-diag.github.io/form/		True
27	1111145627144000	Browser	Chrome	submit	https://bpm-diag.github.io/form/		True
28	1111145627144000	Browser	Chrome	formSubmit	https://bpm-diag.github.io/form/		True

Fig. 7.3 Excerpt of the routine-based log describing 3 out of 50 routine traces of $R_{example}$

different routine variants. For each row, if the corresponding *duplicated* field is set
to *True*, it means the user action associated with that row is present in all the routine
traces of the routine-based log, and the values contained in the data fields mentioned
above are the same across all the routine traces. Otherwise, *duplicated* is set to *False*.

Grouping (line 5)

Once all the rows of the low-level routine-based log have been marked, for each
routine trace, the algorithm evaluates them sequentially (following the timestamped
ordering of events in the trace) and creates different *groups* of events according to
the following conditions:

1. all the sequential rows having the column *duplicated* set to False, that *precede*
   (but are *not preceded* by) a row with the column *duplicated* set to True, are added

A

	case:concept:name	category	application	concept:name	browser_url	tag_value	duplicated
0	1109155433758000	MicrosoftOffice	Microsoft Excel	openWorkbook			True
1	1109155433758000	Browser	Chrome	startDownload	about:blank		True
2	1109155433758000	Clipboard	Clipboard	copy			True
3	1109155433758000	MicrosoftOffice	Microsoft Excel	editCell			True
4	1109155433758000	Browser	Chrome	changeField	https://bpm-diag.github.io/form/	Vanessa Costa	True
9	1109155658364000	MicrosoftOffice	Microsoft Excel	openWorkbook			True
10	1109155658364000	Browser	Chrome	startDownload	about:blank		True
11	1109155658364000	MicrosoftOffice	Microsoft Excel	editCell			True
12	1109155658364000	Clipboard	Clipboard	copy			True
13	1109155658364000	Browser	Chrome	changeField	https://bpm-diag.github.io/form/	Vanessa Marino	True
19	1111145627144000	MicrosoftOffice	Microsoft Excel	openWorkbook			True
20	1111145627144000	Browser	Chrome	startDownload	about:blank		True
21	1111145627144000	MicrosoftOffice	Microsoft Excel	editCell			True
22	1111145627144000	Clipboard	Clipboard	copy			True
23	1111145627144000	Browser	Chrome	changeField	https://bpm-diag.github.io/form/	Paola Valentini	True

B

	case:concept:name	category	application	concept:name	browser_url	tag_value	duplicated
5	1109155433758000	Browser	Chrome	clickRadioButton	https://bpm-diag.github.io/form/	car_no	False
14	1109155658364000	Browser	Chrome	clickRadioButton	https://bpm-diag.github.io/form/	car_yes	False
15	1109155658364000	Browser	Chrome	clickRadioButton	https://bpm-diag.github.io/form/	car_accept	False
24	1111145627144000	Browser	Chrome	clickRadioButton	https://bpm-diag.github.io/form/	car_yes	False
25	1111145627144000	Browser	Chrome	clickRadioButton	https://bpm-diag.github.io/form/	car_reject	False

C

	case:concept:name	category	application	concept:name	browser_url	tag_value	duplicated
6	1109155433758000	Browser	Chrome	clickButton	https://bpm-diag.io/form/		True
7	1109155433758000	Browser	Chrome	submit	https://bpm-diag.io/form/		True
8	1109155433758000	Browser	Chrome	formSubmit	https://bpm-diag.io/form/		True
16	1109155658364000	Browser	Chrome	clickButton	https://bpm-diag.io/form/		True
17	1109155658364000	Browser	Chrome	submit	https://bpm-diag.io/form/		True
18	1109155658364000	Browser	Chrome	formSubmit	https://bpm-diag.io/form/		True
26	1111145627144000	Browser	Chrome	clickButton	https://bpm-diag.io/form/		True
27	1111145627144000	Browser	Chrome	submit	https://bpm-diag.io/form/		True
28	1111145627144000	Browser	Chrome	formSubmit	https://bpm-diag.io/form/		True

Fig. 7.4 Grouping rows of the low-level routine-based log

to a new group. It is worth noticing this condition is satisfied only when a routine trace starts with a sequence of rows having the column *duplicated* set to False;

2. all the sequential rows having the column *duplicated* set to False that *precede* a row with the column *duplicated* set to True (and for which condition 1 does not hold), are added to a new group;

3. all the sequential rows having the column *duplicated* set to True are added to a new group.

In a nutshell, a new group of events will be created for any different sequence of events in a routine trace having the column *duplicated* set to True or False. When the above three steps have been applied for any routine trace in the UI log, the *i-th* groups of each trace will be merged in a larger *i-th* group associated with the UI log. To better understand the rationale of the grouping procedure, let's analyze the routine-based log depicted in Fig. 7.3:

- the sequence of rows [0,4] has the column *duplicated* equals to True, since the user actions associated with that rows are present in all the 3 recorded routine traces, and the values contained in the columns mentioned above are the same across all the routine traces. They violate grouping conditions 1 and 2 but satisfy condition 3. Then, they are added to a group, namely A (cf. Fig. 7.4);
- row 5 has the column *duplicated* equals to False and it follows grouping condition 2, thus it is added to a new group, namely B (cf. Fig. 7.4);
- the sequence of rows [6,8] has the column *duplicated* equals to True for the same reason of the first item. It violates grouping conditions 1 and 2 but satisfies condition 3, thus it is added to a new group, namely C (cf. Fig. 7.4).

The same reasoning can be performed for the other routine traces of the routine-based log. Indeed:

- the sequences of rows [9,13] and [19,23] are added to group A;
- the sequences of rows [14,15] and [24,25] are added to group B;
- the sequences of rows [16,18] and [26,28] are added to group C.

Iterating groups (lines 6-25)

Once all groups have been identified, they are analyzed one by one in a cycle. For each identified group, namely *groupDF* (line 6), if the corresponding column *duplicated* is True for all the rows contained in it (line 7), it means that all the routine traces in that group contain the same user actions. In this case, since that group does not identify a variation point, the rows of the routine trace appearing first in the group (line 8) are directly added to *resultDF* (line 9). Note that choosing the rows of another trace rather than the first one recorded in the group would lead to the same effect. Conversely, if the column *duplicated* is False, it means that we have detected a variation point to be considered.

When the algorithm detects a variation point, it is important to ensure that it is consistent with the routine path executed until that point. Indeed, during each iteration, the rows associated with the previously decided user actions (i.e., those actions selected when a variation point is identified) are saved in *previousDecidedDF* (line 23). A custom routine-based log called *decisionDF* is created to store the rows of the current decision about which user actions enact in the presence of a variation point. In the first cycle iteration, no decision has been made, so *decisionDF* is generated only from the rows of the group that is currently processed (line 18). In the subsequent iterations, the current group and the previous decision are taken into account to find the next possible variation point. The case IDs of the routine traces that have rows in common with the previous decision are selected (line 13). Then, the rows of the routine traces having those case IDs are picked from the current

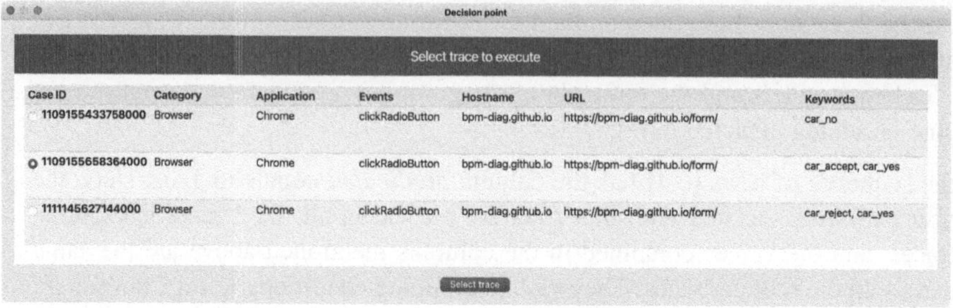

Fig. 7.5 Custom dialog window to enact the user actions of 3 routine traces of $R_{example}$ in the presence of a variation point

group *groupDF* and stored in *filteredDF* (line 14). This step ensures that the next possible variation point is in the routine path that starts from the previously detected variation point.

Finally, *decisionDF* is generated from the rows in *filteredDF* (line 15). In this step, redundant rows are filtered out. Two or more rows of different routine traces are considered as *redundant* if the associated user actions store the same values for the columns mentioned in the step "*Marking*". At this point, the user can choose which user actions to enact in the range of any identified variation point through a custom dialog (e.g., see Fig. 7.5) that is launched just before the script generation step of the approach.

The custom dialog displays data from *decisionDF* (line 20), which is used to bound together all the rows of a routine trace into a single line when a variation point is detected. To better understand this, consider the group of rows with column *duplicated* equals to False in Fig. 7.4 containing 5 rows belonging to 3 different routine traces. The user has to decide which user actions of which routine trace enact, so these 5 rows are grouped together by their ID, and the names of the user actions related to each routine trace are flattened into a single line. Indeed, the dialog in Fig. 7.5 shows a variation point that contains 3 different user inputs that led to 3 different execution variants of $R_{example}$: each line represents a routine trace because it has a unique case ID, and all the user actions names of each routine trace are flattened into the same line.

Once the user decides which user actions to execute (line 21), the corresponding rows are appended to the output routine-based log *resultDF* (line 22). It contains all the rows related to user decisions as well as rows with column *duplicated* equals True (common to every routine trace). Note that *resultDF* will be the input of the Script Generation step of the SmartRPA approach.

Chapter 8
Realizing and Evaluating SmartRPA

This Chapter is structured as follows. Section 8.1 analyzes the architecture and the technical aspects of the tool implementing the SmartRPA approach. Then, we present the results of a multi-step evaluation performed on SmartRPA to investigate the extent to which the proposed approach (and its implemented tool) satisfies four relevant non-functional requirements, namely *robustness/feasibility* (cf. Section 8.2), *effectiveness* (cf. Section 8.3) and *usability* (cf. Section 8.4) employing both synthetic and real-world datasets. The target is to understand if SmartRPA can potentially complement the traditional model-based solutions provided by commercial RPA tools. Section 8.5 concludes the Chapter while Section 8.6 outlines the technical advancements of SmartRPA as of the time of writing this book since the submission of the PhD Thesis.

8.1 Architecture and Development of SmartRPA

Starting from the approach outlined in Fig. 7.2, the architecture of SmartRPA integrates five main SW components developed in Python that enable the reactive synthesis of SW robots according to the intermediate user inputs recorded in the UI logs, thus emulating the most suitable routine variant for any recorded combination of user inputs. An overview of the SmartRPA architecture is shown in Fig. 8.1. The tool can be downloaded and tested at: https://github.com/bpm-diag/smartRPA.

The first SW component of the architecture is an **Action Logger** that concretely implement the *Log Recording* step. The Action Logger provides a Graphical User Interface (GUI) that allows users to select which SW applications s/he wants to record user actions on. All the applications that are not available in the host operating system of the user's computer are disabled by default. Then, the user can start the training session by clicking on the *"Start logger"* button, as shown in Fig. 8.2. The Action Logger provides three categories of logging modules:

© The Author(s), under exclusive license to Springer Nature Switzerland AG 2024
S. Agostinelli: *Generating Executable Robotic Process Automation Scripts from Unsegmented User Interface Logs*, LNBIP 522, pp. 81–97, 2024.
https://doi.org/10.1007/978-3-031-61368-5_8

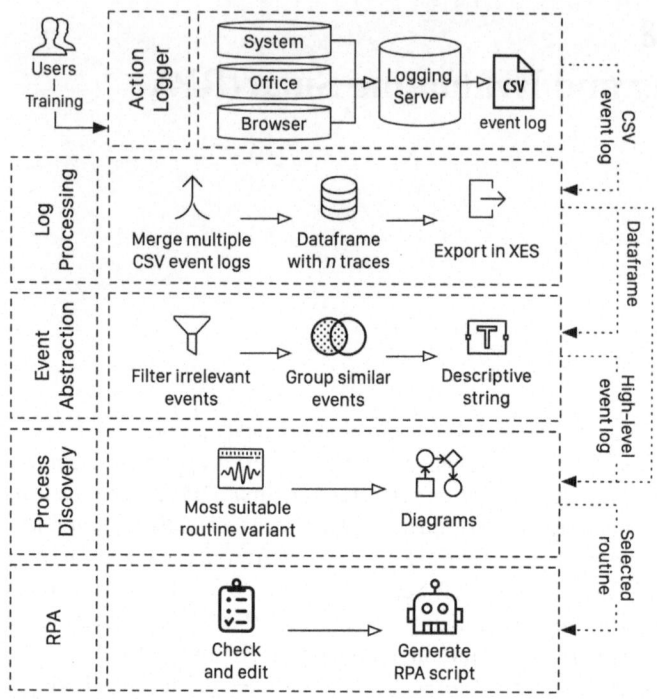

Fig. 8.1 SmartRPA architecture

- *System Logger*: It detects those user actions not related to specific SW applications, i.e.: creation, renaming, movement and deletion of files/folders; copy/paste of files/folders; opening/closing of applications; usage of double-click and hotkeys; insertion/remotion of USB drives.
- *Office Logger*: It detects the user actions performed within Microsoft Office applications, i.e.: Excel, Word, and PowerPoint.
- *Browser Logger*: It detects the user actions performed on web browsers, i.e.: Google Chrome, Mozilla Firefox, Microsoft Edge, and Opera.

Of course, multiple users can run the Action Logger on their computer system many times, performing the same routine in different training sessions. When a training session is completed, i.e., when the routine of interest has been executed from the start to the end, the user can push the *"Stop logger"* button to stop the recording of user actions. The logging modules interact with a Logging Server implemented with the *Flask* framework,[1] which is in charge to store the user actions

---

[1] https://palletsprojects.com/p/flask

(Win) 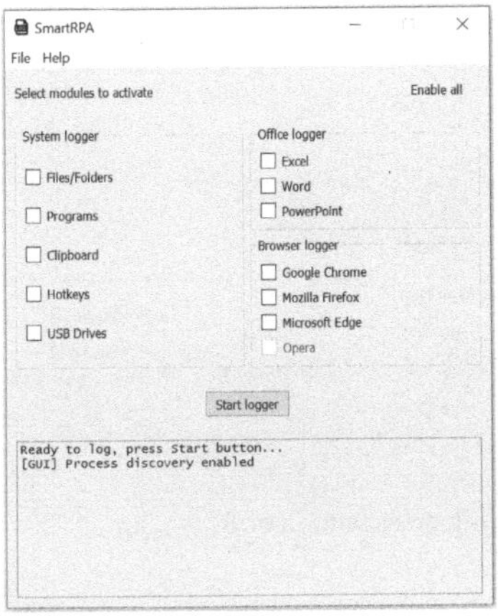   (MacOS)

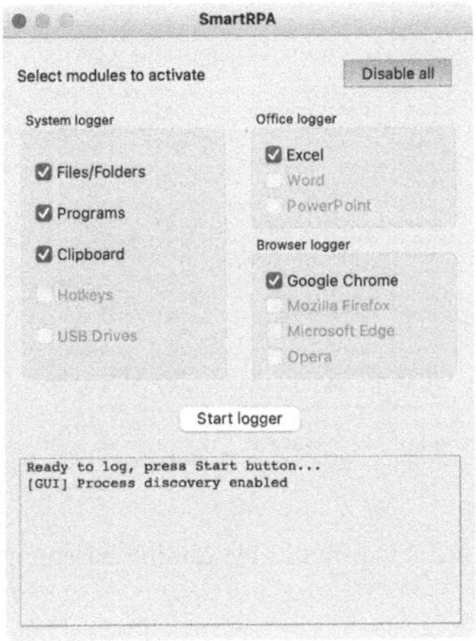

Fig. 8.2 GUI of SmartRPA both on Windows and MacOS

captured by the logging modules and organize them as *events* into several CSV[2] routine-based logs.

Each CSV routine-based log contains exactly one (long) trace of user actions performed in a single training session by a single user. From a technical point of view, *(i)* system events are recorded using different Python modules, including *PythonCOM* (to access the Windows APIs and COM objects like the Microsoft Office suite), and *MacFSEvents* for MacOS; *(ii)* events generated by Microsoft Office applications are recorded using the Office JavaScript APIs; and *(iii)* browser events are recorded using dedicated JavaScript web extensions developed for each supported web browser. In Fig. 8.3, we show a snapshot of a CSV routine-based log recorded in one training session involving the execution of $R_{example}$.

The second SW component of the architecture implements the **Log Processing** step. Specifically, after n training sessions, the Logging Server will deliver the n created CSV routine-based logs to the Log Processing component, which uses Algorithm 3 to import them into a single Pandas dataframe.[3] A dataframe is a two-dimensional size-mutable and heterogeneous tabular data structure with labeled axes (rows and columns), which is used as the main artefact to represent routine-based logs in SmartRPA. Of course, SmartRPA also produces an XES[4] (eXtensible Event Stream) version of the datastream that will contain exactly n traces, one for each recorded CSV routine-based log and can be inspected using the most popular

---

[2] CSV files are file formats that contain plain text values separated by commas. CSV files can be opened by any spreadsheet program, such as Microsoft Excel, Google Sheets, etc. CSV is only capable of storing a single sheet in a file, without any formatting and formulas.

[3] https://pandas.pydata.org/

[4] XES is the standard for the storage, interchange, and analysis of event logs [47]

	A	B	C	D	E	F	G	H	I	J	K	L
1	case:concept:name	case:creator	lifecycle:transition	time:timestamp	org:resource	category	application	concept:name	event_src_path	event_dest_path	clipboard_content	mouse_coord
2	1,10916E+15	SmartRPA	by complete	2020-11-09T15:5!	marco	Browser	Chrome	startDownload				
3	1,10916E+15	SmartRPA	by complete	2020-11-09T15:5!	marco	MicrosoftOffice	Microsoft Excel	openWorkbook	/Users/marco/desktop/travel authorization request procedure.xlsx			
4	1,10916E+15	SmartRPA	by complete	2020-11-09T15:5!	marco	Browser	Chrome	reload				
5	1,10916E+15	SmartRPA	by complete	2020-11-09T15:5!	marco	Clipboard	Clipboard	copy			Vanessa Costa	
6	1,10916E+15	SmartRPA	by complete	2020-11-09T15:5!	marco	MicrosoftOffice	Microsoft Excel	editCell	/Users/marco/desktop/travel authorization request procedure.xlsx			
7	1,10916E+15	SmartRPA	by complete	2020-11-09T15:5!	marco	Browser	Chrome	clickTextField				366,4
8	1,10916E+15	SmartRPA	by complete	2020-11-09T15:5!	marco	Browser	Chrome	paste			Vanessa Costa	
9	1,10916E+15	SmartRPA	by complete	2020-11-09T15:5!	marco	Browser	Chrome	changeField				
10	1,10916E+15	SmartRPA	by complete	2020-11-09T15:5!	marco	Clipboard	Clipboard	copy			Full professor	
11	1,10916E+15	SmartRPA	by complete	2020-11-09T15:5!	marco	MicrosoftOffice	Microsoft Excel	editCell	/Users/marco/desktop/travel authorization request procedure.xlsx			346,498
12	1,10916E+15	SmartRPA	by complete	2020-11-09T15:5!	marco	Browser	Chrome	clickTextField				
13	1,10916E+15	SmartRPA	by complete	2020-11-09T15:5!	marco	Browser	Chrome	paste			Full professor	
14	1,10916E+15	SmartRPA	by complete	2020-11-09T15:5!	marco	Browser	Chrome	changeField				
15	1,10916E+15	SmartRPA	by complete	2020-11-09T15:5!	marco	Clipboard	Clipboard	copy			ds_mail@uniroma1.it	
16	1,10916E+15	SmartRPA	by complete	2020-11-09T15:5!	marco	MicrosoftOffice	Microsoft Excel	editCell	/Users/marco/desktop/travel authorization request procedure.xlsx			256,579
17	1,10916E+15	SmartRPA	by complete	2020-11-09T15:5!	marco	Browser	Chrome	mouseClick				
18	1,10916E+15	SmartRPA	by complete	2020-11-09T15:5!	marco	Browser	Chrome	paste			ds_mail@uniroma1.it	
19	1,10916E+15	SmartRPA	by complete	2020-11-09T15:5!	marco	Browser	Chrome	changeField				
20	1,10916E+15	SmartRPA	by complete	2020-11-09T15:5!	marco	Clipboard	Clipboard	copy			CNTC5T19A58B1765	
21	1,10916E+15	SmartRPA	by complete	2020-11-09T15:5!	marco	MicrosoftOffice	Microsoft Excel	editCell	/Users/marco/desktop/travel authorization request procedure.xlsx			
22	1,10916E+15	SmartRPA	by complete	2020-11-09T15:5!	marco	Browser	Chrome	clickTextField				329,379
23	1,10916E+15	SmartRPA	by complete	2020-11-09T15:5!	marco	Browser	Chrome	clickTextField				329,379
24	1,10916E+15	SmartRPA	by complete	2020-11-09T15:5!	marco	Browser	Chrome	doubleClick				

Fig. 8.3 Snapshot of the routine-based log captured during an execution of $R_{example}$

process mining tools, such as *ProM*,[5] *Disco*[6] or *Apromore*.[7] The dataframe created by Algorithm 3 consists of low-level events with fine granularity associated one-by-one to a recorded user action (e.g., mouse clicks, file selections, etc.). Each row of the dataframe includes 45 columns with relevant data about the recorded event, i.e., its payload, such as: the timestamp, the application that generated the event, the resources involved, etc., cf. Fig. 8.3.

At this point, an **Event Abstraction** component is used to produce a high-level routine-based log from the low-level one, by performing the following steps:

1. *Filtering noise/irrelevant events.* The Action Logger records many low-level events in the dataframe-based routine-based log, such as the interaction with

---

Algorithm 3: Processing CSV routine-based logs

Parameters: A list of logs `fileList`
Result: A DataFrame `combinedDF`, A XES file `logXES`

1   *createDirectories*();   // where files will be saved
2   dfs ← *list*();   // list of dataframes
3   **for** *any CSV log in* `fileList` **do**
4   df ← import CSV log into a DataFrame;
5   df ← rename columns to match XES standard;
6   df ← sort rows by timestamp;
7   df ← create `case:concept:name` column based on the first timestamp;
8   dfs.append(df);
9   **end**
10   combinedDF ← combine all dataframes in dfs into a single one;
11   logXES ← *export*(combinedDF);   // exported as XES file
12   **return** (combinedDF, logXES)

---

[5] http://www.promtools.org/

[6] https://fluxicon.com/disco/

[7] https://apromore.org/

Algorithm 4: Event Abstraction

Parameters: A DataFrame `df`
Result: A DataFrame `HighLevelDF`
1 `HighLevelDF` = *None*;
2 `df` ← filter irrelevant rows from `df`;
3 `df` ← group similar events in `df`;
4 **for** *row in `df`* **do**
5   `descriptiveRow` ← create descriptive string from row;
6   `HighLevelDF`.append(`descriptiveRow`);
7 **end**
8 **return** `HighLevelDF`

---

the browser windows (e.g., user actions "resize", "open", "close"), tabs (e.g., user actions "move", "open", "close") and content (page zoom, installing extensions). From a workflow perspective, these events are not relevant for any RPA analyst that aims to understand the general behaviour of the routine. For this reason, they are filtered out by the high-level routine-based log under construction.

2. *Grouping similar events.* Within a dataframe-based routine-based log, different low-level events can refer to the same high-level concept. For example, in a web page, the Action Logger can capture 7 different types of clicks, based on the element that's being clicked ("clickButton", "clickTextField", "doubleClick", "clickTextField", "mouseClick", "clickCheckboxButton", "clickRadioButton"). All these events just indicate that the user, during the training session, has clicked on an interactive element on the UI, thus the high-level workflow of the routine may just show the action "Click on button", because from the RPA analyst perspective, it is not relevant what kind of click was performed.

3. *Creating descriptive labels.* Any recorded event provides a low-level description of the nature of the user action performed. For example, if the user edits a cell in Excel, the Action Logger records one of these events: "editCellSheet", "editCell", or "editRange". From the RPA analyst perspective, all such events refer to the same concept of "Editing a cell". To this aim, to make the user action underlying an event more descriptive for the RPA analyst, further information (stored in the low-level dataframe-based routine-based log) can be added to its label, such as the cell and the sheet edited, the value inserted, etc. This allows us to create a (more) descriptive label for any event in the high-level routine-based log, e.g., *"Edit cell B3 on Sheet 1 with value 'x'"*.

Concretely, the Event Abstraction component is realized enacting the above steps through Algorithm 4, and the outcome will be a high-level routine-based log to be used by the next component of the architecture.

At this point, the **Process Discovery** component of the architecture comes into play. Starting from the high-level routine-based log generated by the Event Abstraction component, it applies the heuristic miner algorithm (the decision to employ the heuristic miner has been driven by its ability to discover highly understandable flowcharts from a BPM analyst perspective [7]) implemented in PM4PY [20] to

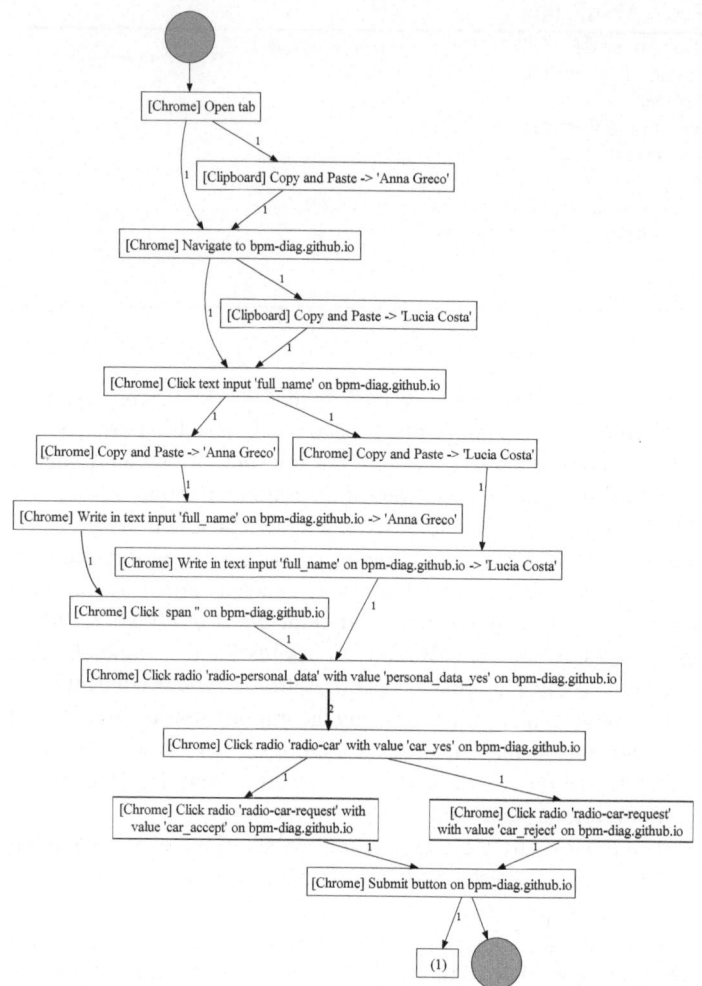

Fig. 8.4 DFG describing a portion of the high-level workflow of $R_{example}$

derive the high-level workflow describing the overall users' observed behaviour as a Directly-Follows Graph (DFG). We show in Fig. 8.4 a portion of the high-level workflow discovered from the high-level routine-based log associated with $R_{example}$.

Then, it applies Algorithm 2 (described in detail in Section 7.4) to automatically detect the different routine variants among all the routine traces stored in the low-level dataframe-based routine-based log, by evaluating the potential intermediate user inputs required to emulate the most suitable version of the routine on the UI.

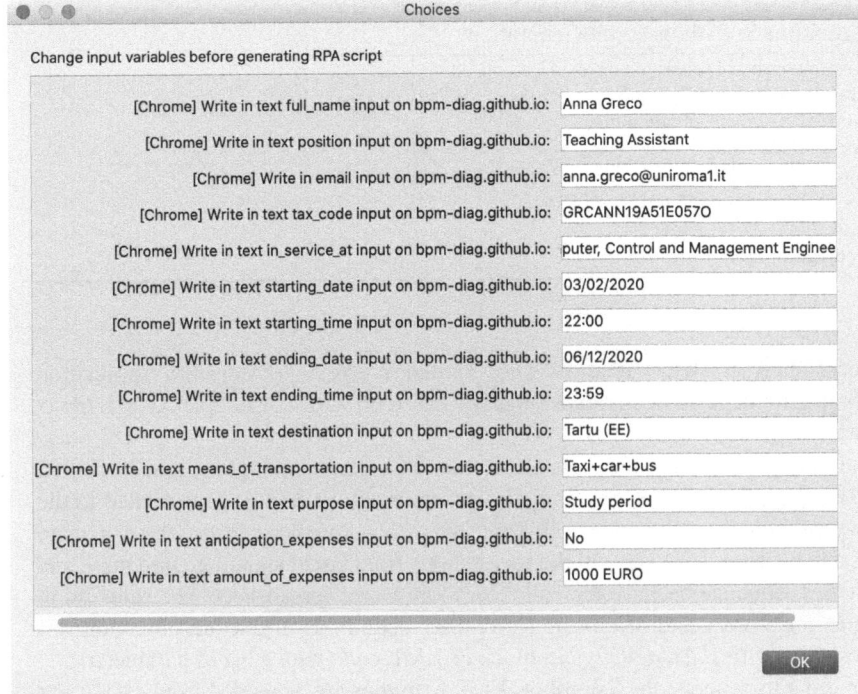

Fig. 8.5 Custom dialog window to personalize editable fields of $R_{example}$

Finally, there is the **Script Generation** component. Once the routine variant to automatize is selected, before its enactment with a SW robot, an RPA analyst can personalize the values stored in its events through a custom dialog window (cf. Fig. 8.5).

The tool automatically detects the events that can be edited, such as typing something on a web page, renaming a file, pasting a text, or editing an Excel cell, and dynamically builds the GUI to let the RPA analyst edit them. After confirmation, the low-level dataframe-based routine-based log is updated. Finally, the Python executable script based on the selected RPA routine and updated with the RPA analyst's edits, is generated by scanning the recorded low-level events in the dataframe-based log and converting them into executable pieces of SW code in Python, through Algorithm 5. To properly work the script generation algorithm relies on *Automagica*,[8] an Open Source framework for process automation, and *Selenium*,[9] a popular suite of tools for automating web browsers.

SmartRPA is also able to generate executable RPA scripts compatible with *UiPath*, a tool that allows visually designing automation processes. Once the routine variant to automate along with the RPA analyst's edits has been generated, the UiPath script

---

[8] https://github.com/automagica/automagica

[9] https://www.selenium.dev/

Algorithm 5: Python Script Generation

Parameters: A DataFrame df
Result: A Python SW Robot script
1 script ← create a Python file;
2 **for** *row in df* **do**
3 | pythonEvent ← *generatePythonEvent*(row);
4 | append pythonEvent to script;
5 **end**
6 **return** script

---

is accordingly created. UiPath files are written in XAML (Extensible Application Markup Language), a declarative language based on XML. A sample XAML file is shown in Fig. 8.6.

It is composed by a *Main Sequence* containing, in turn, multiple sequences, based on the category of the user actions. For example, all the user actions related to the browser should be wrapped by a *Browser Activities* sequence because they all share the same browser. Likewise, all the user actions from Excel should go into the *Excel Activities* sequence because they all refer to an Excel spreadsheet. The same thing applies for *System* and *Microsoft Office* user actions. Every sequence contains a series of activities. An *activity* is a block of XML code with a list of parameters.

In order to generate the SW robot, XML activities are generated from each event in the low-level dataframe-based routine-based log using lxml[10] Python library. Activities are created with Python methods which take parameters as input and

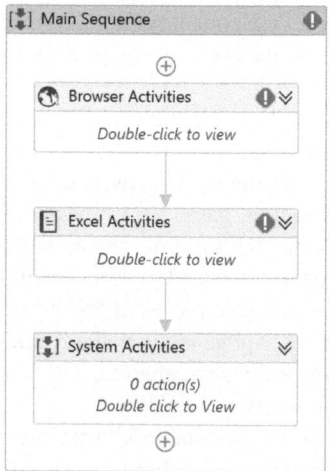

Fig. 8.6 UiPath sequence

---

[10] https://lxml.de/

Algorithm 6: UiPath Script Generation

Parameters: A DataFrame df
Result: A UiPath SW Robot mainSequence
1 mainSequence ← create main XML sequence;
2 activities ← *dict*();   // dictionary to store XML activities
3 lastIndex ← *False*;   // True in the last loop iteration
4 categoryChange ← *False*;   // True when there is a category change
5 **for** *row in df* **do**
6 | XMLNode ← *generateXMLNode*(row);
7 | activities[currentCategory].*append*(XMLNode);
8 | **if** *categoryChange or lastIndex* **then**
9 | | browserSeq ← *createSequence*(activities['Browser']);
10 | | mainSequence.*append*(browserSeq);
11 | | MSOfficeSeq ← *createSequence*(activities['MicrosoftOffice']);
12 | | mainSequence.*append*(MSOfficeSeq);
13 | | OpSysSeq ← *createSequence*(activities['OperatingSystem']);
14 | | mainSequence.*append*(OpSysSeq);
15 | | activities.*clear*();   // empty dictionary
16 | **end**
17 **end**
18 write mainSequence to XAML file;   // UiPath Project
19 **return** mainSequence

---

return XML nodes, as shown in Algorithm 6. The algorithm describes generating a UiPath SW robot from a given low-level dataframe-based routine-based log.

The main XML sequence that will contain all other sequences is created, along with a dictionary to store activities based on their category (*browser, office, system*). For each event in the low-level dataframe-based routine-based log, a corresponding XML node is generated and appended to the dictionary based on its category. When there is a change in category and in the last loop iteration, user actions are wrapped in a sequence-specific for that category and added to the main sequence. The *activities* dictionary is cleared before restarting the loop to prevent duplicate activities in the main sequence. Finally, the generated XML sequence is written as a XAML file that can be opened and run in UiPath.

A screencast with installation instructions showing the working of SmartRPA is available in the Github repository of the tool at: https://github.com/bpm-diag/smartRPA/.

8.2 Assessing the Robustness and Feasibility of the Algorithm for the Automated Detection of Variation Points

To investigate the *robustness* and *feasibility* of our approach to the reactive synthesis of SW robots from UI logs, we performed several synthetic experiments employing UI logs of increasing complexity. Specifically, we generated 240 different UI logs (containing in total 150.000 different routine traces), in a way that each UI log was characterized through a unique configuration obtained by varying the following input settings:

- *log_size*: number of traces in the UI log (250/500/750/1000);
- *trace_size*: number of events in each routine trace (25/50/75/100);
- *events_size*: number of possible different events to be considered for the creation of a trace (40/80/120);
- *variation_points*: number of different variation points included in the UI log (1/2/3/4/5).

Note that the number of possible decisions to be taken in a variation point was generated randomly, ranging from 2 to 10 possible outgoing decisions. Following our definition of variation points explained in Section 2.4, each UI log was generated creating log traces having a similar structure in terms of recorded events, except for the presence of the variation points. Repeated events and concurrency are allowed inside a UI log. However, since they have been randomly introduced in UI logs, we can not provide solid findings related to their impact on the identification of the variation points. The synthetic UI logs generated for the test are available for testing and experiments repeatability at: https://tinyurl.com/yyk68psx.

The target was to investigate if the amount and anatomy of variation points discovered by SmartRPA are the same that was synthetically introduced in the sample routine executions recorded in the UI logs (i.e., *robustness*), and to measure the performance of the entire approach to generate a SW robot by solely using the UI logs (i.e., *feasibility*).

Concerning the robustness of the approach, for all the 240 tested logs the approach was able to always discover the correct variation points to be considered for the synthesis of SW robots. It is worth noticing that this result is justified by the fact that we employed a fixed (yet large) alphabet of user actions for the generation of the sample UI logs, in line with the assumption that a routine reflects highly predictable and repetitive work with low flexibility requirements (and, consequently, with a low number of variants) [31]. Such an assumption is remarkably realistic for SmartRPA, as the UI logs used for the reactive synthesis of SW robots are recorded during controlled training sessions in which many users are instructed to perform any time the same routine. On the other hand, in the case of more flexible procedures, our algorithm would detect a new variation point any time there are distinct user actions (cf. Section 7.4) recorded at the same point of different executions of the same routine. This would lead to a consistent growth of the amount of identified variation points, which is not wrong in principle, but that could not be suitable to concisely represent the behaviour of a routine. Therefore, we can state that our algorithm for the detection of variation points is *robust* if the UI logs have the features outlined in the experiment settings. In the absence of further experiments, we can not state anything about the robustness of the algorithm when our working assumptions are contradicted, i.e., when more flexible (i.e., non-repetitive) procedures are executed. Note that the literature proposes dedicated approaches to detect the decision points in case of less flexible processes to be analyzed, e.g., see [28], even if the granularity

Table 8.1 Experimental results showing the *feasibility* of SmartRPA to the reactive generation of SW robots (only logs with 1000 traces are shown here). The time (in *milliseconds*) is the average per trace.

Event size: 40	Time				
Trace size	1	2	3	4	5
25	0.453	0.452	0.53	0.409	0.423
50	0.417	0.433	0.417	0.425	0.419
75	0.439	0.511	0.424	0.43	0.431
100	0.454	0.416	0.421	0.424	0.431

Event size: 80	Time				
Trace size	1	2	3	4	5
25	0.422	0.428	0.43	0.413	0.412
50	0.427	0.425	0.444	0.417	0.428
75	0.42	0.428	0.553	0.422	0.437
100	0.442	0.434	0.428	0.438	0.432

Event size: 120	Time				
Trace size	1	2	3	4	5
25	0.413	0.507	0.421	0.416	0.421
50	0.421	0.412	0.417	0.42	0.421
75	0.425	0.433	0.438	0.451	0.429
100	0.437	0.433	0.428	0.532	0.523

of the process activities is less fine than the one of the user actions involved in a routine execution.

The feasibility was measured in terms of the computation time required to generate a SW robot starting from UI logs of growing complexity. The results, which are summarized in Table 8.1, indicate that the total computation time increases with the number of traces in the UI log,[11] ranging from ~ 100ms for UI logs with 250 traces up to ~ 500ms for event logs with 1000 traces. This result was expected, since more traces in a UI log mean more executions to analyze and interpret. On the other hand, if we consider a fixed log size, it seems that the performance of the approach scales very well in case of an increasing number of variation points to be discovered and log traces/alphabet of events of growing size. Sometimes, it has been also observed that SmartRPA gets faster by adding events in a trace, which suggests that the performance of the approach does not suffer the presence of a larger alphabet of events.

---

[11] For the sake of readability, the table includes only the results related to UI logs containing 1000 traces. The complete list of results can be analyzed at: https://tinyurl.com/y55v56qa

8.3 Evaluating the Effectiveness of SmartRPA

To address **RQ2.3**, we enacted a controlled experiment involving real users exploiting the use case of Section 2.6 to investigate the effectiveness of the SmartRPA approach when compared to UiPath, which is one of the major vendors in the RPA market according to [13], and realizes the "traditional" model-based approach for the generation of SW robots.

To this end, we conducted a user study based on the use case presented in Section 2.6, by asking 20 different administration employees to fill the Google Form using the data from the Excel spreadsheet containing the information to apply for a travel request. All the user actions were enacted on distinct computer systems having different features and operating systems. During the execution of their routine, the employees were coupled with a first group of 20 (out of a sample of 40) Master students of the course of Process Management and Mining (PMM) held at Sapienza University of Rome (one student per employee), which were requested to observe the execution steps of $R_{example}$. We denote with p_{B1} this first group of users. In parallel, a second group of 20 users was remotely connected to the employees' computer systems, with the target to record the user actions performed on the UI of such systems exploiting the Action Logger component of SmartRPA, thus generating at the end 20 different UI logs. We denote with p_{B2} this second group of users. It is worth noticing that all the PMM students involved in the user study can be considered as *expert users* in business process modeling and automation.

At this point, we requested any of the 20 expert users in p_{B1} to employ UiPath to model a flowchart diagram associated with $R_{example}$ and generate the associated SW robot using the functionalities of the UiPath framework. On the other hand, we asked any of the 20 expert users in p_2 to exploit the UI logs storing the executions of $R_{example}$ as inputs to use SmartRPA for the generation of the associated SW robot.

To assess the effectiveness of SmartRPA to synthesize SW robots from UI logs, we investigated the following experimental hypothesis $\mathbf{H}_{B1}$: *Employing the SmartRPA approach, thus neglecting the manual specification stage of the routine behaviour, is more effective than employing traditional approaches that require to manually specify and implement the behaviour of SW robots by means of flowchart models.* To this aim, we have first built the null hypothesis $\mathbf{H}_{B0}$: *Employing the SmartRPA approach does not provide any advantage in terms of effectiveness if compared with traditional modeling-driven RPA approaches.* Then, to support or reject $\mathbf{H}_{B0}$, a *between-subject approach* was used, i.e., each user in p_{B1} (p_{B2}, respectively) was assigned to a different experimental condition, related to the exclusive use of UiPath (c_{B1}) or SmartRPA (c_{B2}) to perform the required steps for the generation of the SW robot for $R_{example}$. Any user in p_{B1} (p_{B2}, respectively) was preliminarily instructed about the functionalities of UiPath (SmartRPA, respectively) through a short training session. Notice that we selected users that were completely unaware about the use of both UiPath and SmartRPA before the starting of the experiment.

We evaluated the validity of $\mathbf{H}_{B0}$ by asking any expert user that completed the user study the following three questions:

- Q_{B1}: The development life-cycle of a SW robot (from the definition of the routine behaviour to the generation and execution of the associated SW robot) is a time-consuming task. Do you agree?
- Q_{B2}: The extraction of the routine's knowledge required for the development and execution of a SW robot is a complex task. Do you agree?
- Q_{B3}: Once a SW robot has been generated, the monitoring of its execution and the inspection of its behaviour is a complex task. Do you agree?

Questions are rated with a 7-point average numerical scale structured as follows: 1 ("Strongly Disagree"), 2 ("Disagree"), 3 ("Somewhat Disagree"), 4 ("Neither Agree nor Disagree"), 5 ("Somewhat Agree"), 6 ("Agree"), 7 ("Strongly Agree"). We kept the same difference between subsequent points of the scale, as suggested by [34]. The choice to employ a 7-point scale (rather than a 5-point scale) is supported by the findings of Sauro [90], which states that in case of a questionnaire consisting of few questions *"having seven points tends to be a good balance between having enough points of discrimination without having to maintain too many response options"*.

To evaluate the answers associated to Q_{B1}, Q_{B2} and Q_{B3} we performed a comparison of the rates obtained from the questionnaire, respectively in the cases of c_{B1} and c_{B2}. Specifically, for each question, we employed a *2-Sample t-test* with a 95% confidence level to determine whether the means between the two distinct populations (i.e., independent groups p_{B1} and p_{B2}) involved in c_{B1} and c_{B2} differ. Before running the 2-Sample t-test, we first exploited the Kolmogorov Smirnov Statistic (KS Test) to establish the normality of the distribution of the collected data [26], and then we checked that the variances and standard deviations in both groups were approximately equal [90].

Finally, we measured the level of statistical significance by analyzing the resulting *p-value*. We remind that a p-value ≤ 0.05 is considered to be statistically significant, while a p-value ≤ 0.01 indicates that there is substantial evidence in favour of the experimental hypothesis. In addition, because the result obtained by each question generates its own test statistic, we applied the Benjamini-Hochberg False Discovery Rate correction for multiple testing [19] controlling for a false discovery rate of 0.25, which seems a reasonable cut-off given the novelty of our effectiveness experiments in the field of RPA. The results of the analysis are summarized in Table 8.2.

It appears evident that the null hypothesis $\mathbf{H}_{B0}$ is statistically supported by the results obtained for Q_{B3}, while it is rejected for Q_{B1} and Q_{B2}. Concerning Q_{B3}, there is strong evidence that a traditional model-based approach based on designing routines by means of flowchart diagrams (like UiPath) is more effective to monitor the behaviour of the (running) SW robots associated with the routines and inspecting the related RPA scripts. On the other hand, to skip the modeling task entirely by employing an approach based only on UI logs enables a faster generation of SW robots (cf. Q_{B1}) requiring solely the knowledge stored in the UI logs (cf. Q_{B2}). In summary, we can conclude that log-based approaches like SmartRPA increase the degree of automation of the design-time steps required to generate SW robots, reducing the intervention of human experts in this phase. Therefore, $\mathbf{H}_{B1}$ can be considered as validated for Q_{B1} and Q_{B2} but rejected for Q_{B3}, where model-based approaches appear to be more effective to monitor running SW robots.

Table 8.2 Effectiveness of SmartRPA: p-values associated to each question

Q_{B1}		Q_{B2}		Q_{B3}	
SmartRPA	UiPath	SmartRPA	UiPath	SmartRPA	UiPath
2	3	2	3	3	1
2	3	2	3	3	2
3	3	3	3	3	2
3	3	3	3	3	2
3	4	3	3	3	3
3	4	3	4	3	3
3	4	3	4	4	3
3	4	3	4	4	3
4	4	3	4	4	3
4	4	4	4	4	3
4	4	4	4	4	3
4	4	4	4	4	4
4	4	4	4	4	4
4	4	4	4	4	4
4	5	4	5	4	4
4	5	4	5	4	4
4	5	4	5	5	4
4	5	4	6	5	4
5	6	5	6	6	4
5	6	5	6	6	4
p-value: 0.0333436		p-value: 0.0317368		p-value: 0.0081553	

8.4 Quantifying the Usability of the UI of SmartRPA

Last but not least, we investigated the degree of *usability* of the UI developed for SmartRPA. Specifically, we administered the SUS questionnaire [23] to the 20 expert users that were involved in the experimental condition c_{B2}, i.e., that used SmartRPA. The questionnaire consists of 10 statements evaluated with a Likert scale that ranges from 1 ("strongly disagree") to 5 ("strongly agree") adapted to SmartRPA:

- I think that I would like to use SmartRPA frequently.
- I found SmartRPA unnecessary complex.
- I thought SmartRPA was easy to use.
- I think that I would need the support of a technical person to be able to use SmartRPA.
- I found the various functions in SmartRPA well integrated.
- I thought there was too much inconsistency in SmartRPA.
- I would imagine that most people would learn to use SmartRPA very quickly.
- I found SmartRPA very awkward to use.
- I felt very confident using SmartRPA.
- I needed to learn a lot of things before I could get going with SmartRPA.

At the end of the questionnaire, an overall score is assigned to the questionnaire. The score can be compared with several benchmarks presented in the research literature to determine the usability of the tool being evaluated. In our test, we made

Table 8.3 Overview of SUS results.

User	q1	q2	q3	q4	q5	q6	q7	q8	q9	q10	SUS Score	Average
p1	4	2	4	2	4	2	4	2	4	2	75.0	79.3
p2	4	1	4	2	4	2	3	2	4	2	75.0	
p3	3	1	4	1	4	2	5	1	1	1	77.5	
p4	4	1	1	1	4	2	4	2	4	1	75.0	
p5	3	1	5	1	4	2	4	2	4	1	82.5	
p6	4	2	5	2	4	1	4	2	5	1	85.0	
p7	1	1	5	1	4	2	5	1	5	1	85.0	
p8	3	2	4	2	4	2	4	2	4	2	72.5	
p9	4	2	5	1	4	2	4	1	4	1	85.0	
p10	4	2	4	2	4	2	4	2	5	2	77.5	
p11	4	2	4	2	4	2	4	2	4	2	75.0	
p12	4	2	4	2	4	2	4	2	4	2	75.0	
p13	3	2	4	2	4	2	4	2	4	2	72.5	
p14	4	1	4	2	4	2	4	2	4	2	77.5	
p15	4	2	2	2	3	1	4	2	4	1	72.5	
p16	4	1	5	2	4	2	5	2	4	2	82.5	
p17	4	2	4	1	4	3	4	1	4	2	77.5	
p18	4	2	4	1	3	3	4	1	4	1	77.5	
p19	4	1	5	1	5	1	5	1	4	2	92.5	
p20	5	2	5	2	4	1	5	1	5	1	92.5	

use of the benchmark given in [90], which associates to each range of the SUS score a percentile ranking varying from 0 to 100, indicating how well it compares to other 5,000 SUS observations performed in the literature.

The collection of the ranks associated with any statement of the SUS is reported in Table 8.3, calculated following the steps discussed in [90]. Since the average SUS score obtained by the tool was 79.3, according to the selected benchmark [90], the usability of the tool corresponds to a rank of A-, which indicates a degree of usability among very good and excellent.

8.5 Threats to Validity

A series of common issues may influence the results of our evaluation, such as the (random) selection of the sample of users who performed the experiments (even if from a well-defined population, which mitigates the issue), the selection of the statistical tests to evaluate the collected data, etc.

Notably, while the controlled experiments employed to measure the effectiveness of SmartRPA appears to have an high internal validity due to the control of the experimental conditions exercised throughout the experiment, on the other hand, this control can cause the experiment to have a questionable external validity. This is due to the complexity to replicate the experimental conditions in real-world settings that have many extraneous variables at play, making the findings less generalizable.

(Win) 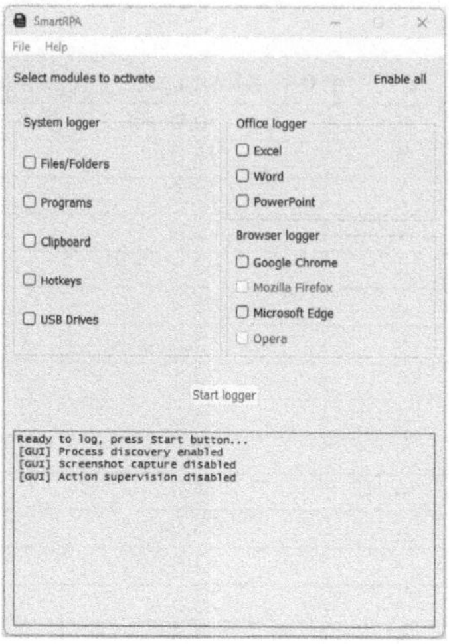   (MacOS)

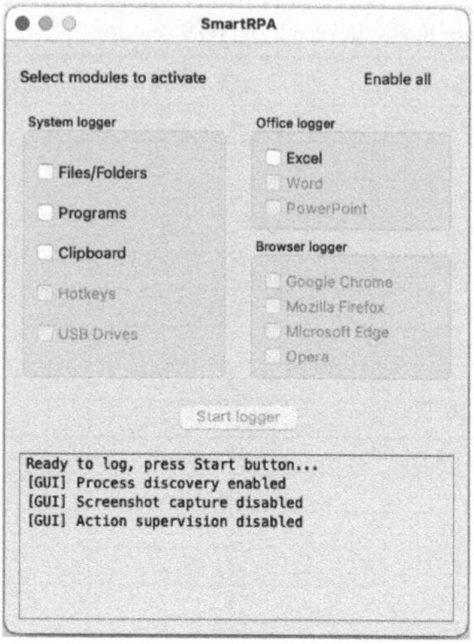

Fig. 8.7 GUI of SmartRPA both on Windows and MacOS

However, we observe that we do not claim that our results are representative of all RPA literature, or to be generalizable to other fields or contexts.

Concerning the experiments' findings, we claim that their validity is bound to the experiments settings. For example, in the case of the experiment to measure the effectiveness of SmartRPA, using a 2-Sample t-test with a 95% confidence level enables us to state that we are 95% confident that the null hypothesis $\mathbf{H}_{B0}$ is partially rejected. However, performing a further experiment that includes more users and the application of a second confidence level (e.g., set to 99%) could support more substantial evidence of the results.

8.6 SmartRPA Advancements

Between the time of the PhD Defense in May 2022 and the writing of this book, SmartRPA has been upgraded primarily from a technical point of view. Even though the GUI remains largely the same, the logging component has been extended, as depicted in Fig. 8.7.

Specifically, each event generated by the logging modules is now correlated with a screenshot of the UI where the event has been generated. Mining screenshot data alongside the logging modules facilities can be valuable for generating translucent event logs [21]. By analyzing screenshots of user interactions, analysts can identify potential paths or human reasoning beyond the recorded event sequence, providing a more complete picture of the process [75, 76]. To further enhance efficiency, users can capture feedback directly after each action using an annotation feature. This

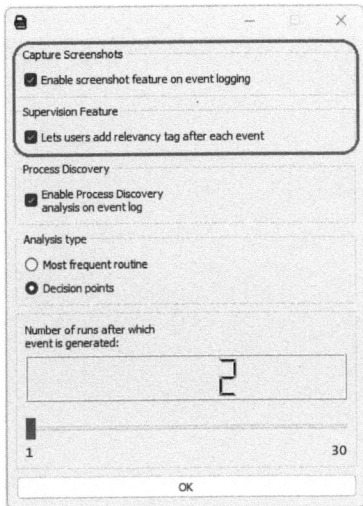

Fig. 8.8 SmartRPA preferences

feature allows users to tag actions as necessary or not, streamlining the SmartRPA filtering process and reducing the need for developer supervision to create a better return on investment in the RPA implementation process [78]. Screenshots and supervision features are disabled by default, but they can be activated by clicking on the preferences menu as depicted in Fig. 8.8.

Part IV
Conclusion

Chapter 9
Limitations, Future Works and Impact

RPA recently gained a lot of attention in the BPM domain [103]. Since RPA operates at the UI level, rather than at the system level, it allows applying automation without any changes in the underlying information system. Thus, the entry barrier of adopting RPA in BPs that are already in place is lower compared to conventional BPM [40]. However, the current generation of RPA tools is driven by predefined rules and manual configurations made by expert users rather than by automated techniques [69], preventing the widespread adoption of these tools in the BPM domain.

Still, to date, a great deal of time is required to identify the routines for automation and manually program the SW robots. Even if RPA tools are able to automate a wide range of routines, they cannot determine which routines should be automated in the first place. Indeed, in the early stages of the RPA life-cycle it is required to: *(i)* identify the candidate routines to automate through interviews and detailed observation of workers conducting their daily work, *(ii)* record the interactions that take place during the routines' enactment on the UI of software applications into dedicated UI logs, and *(iii)* manually specify their conceptual and technical structure (often in form of flowchart diagrams) for identifying the behaviour of SW robots. Towards this direction, the presented book tries to mitigate the involvement of skilled human experts by delivering two contributions to the RPA community:

1. An interactive approach to the automated segmentation of UI logs (**C1**).
2. The SmartRPA approach to the automated identification of the variation points of a routine, to enable the selection of the most suitable routine variants to be implemented with a SW robot directly from a routine-based log (**C2**), the output of the segmentation task.

In Section 9.1 and Section 9.2 we summarize the limitations, impact and future work of the main contributions of this book. Finally, Section 9.3 provides some final considerations that apply generally to the overall research underlying this book.

S. Agostinelli: *Generating Executable Robotic Process Automation Scripts from Unsegmented User Interface Logs*, LNBIP 522, pp. 101–104, 2024.
https://doi.org/10.1007/978-3-031-61368-5_9

9.1 Overview on the Issue of Automated Segmentation of UI Logs

For tackling **C1**, this book presented an approach that relies on three main steps: *(i)* a frequent-pattern identification technique (customized on an ad-hoc basis) to automatically derive the routine segments from a UI log, *(ii)* a human-in-the-loop interaction to filter out those segments not allowed (i.e., wrongly discovered from the UI log) by any real-world routine execution, and *(iii)* a routine traces detection component that leverages trace alignment in Process Mining to cluster all user actions belonging to a specific routine segment into well-bounded routine traces. Our approach is based on a semi-supervised assumption, since we know a priori the end-delimiters to be associated with any user action that ends a routine execution. On the other hand, the approach is not aware of the concrete behaviour of the routines of interest, which will be discovered by the approach itself. For this reason, we consider this contribution as an important step towards the development of a more complete and unsupervised technique to the segmentation of UI logs.

The presented approach is able to extract routine traces from unsegmented UI logs that record in an interleaved fashion many different routines but not the routine executions, thus losing in accuracy when there is the presence of interleaving executions of the same routine. In addition, it is also able to properly deal with shared user actions required by all routine executions in the UI log, thus achieving the cases *1.1*, *2.1*, *2.3*, *3.1*, and *3.3*. It is worth noticing that the routine traces detection component can be employed as a stand-alone supervised segmentation technique [11] able to achieve all variants of cases *1*, *2*, and (partially) *3*, except when there are interleaved executions of shared user actions of many routines. In that case, the risk exists that a shared user action is associated with a wrong routine execution (i.e., Case *3.3* and Case *3.4* are not covered). The *supervised* assumption, which consists of knowing a priori the structure of routines (i.e., the interaction models), may ease the segmentation task. Still, as a side effect, it may strongly constrain the discovery of routine traces only to the "paths" allowed by the routines' structure, thus neglecting that some valid yet infrequent routine variants may exist in the UI log. For this reason, the novelty of the proposed approach to the segmentation of UI logs [3] is to semi-automatically discover such structures in the form of routine segments and then use them as input for the routine traces detection component [11].

As an immediate future work, we are going to perform a more robust evaluation on real-world case studies with heterogeneous UI logs, and we aim at relaxing the semi-supervised assumption by employing *machine learning* techniques to the au-tomated identification of the end-delimiters. This work could also play a crucial role in the context of AI-augmented Business Process Management Systems (ABPMSs), that is, an emerging class of process-aware information systems empowered by AI technology for autonomously unfolding and adapting the BP execution flows [35]. Indeed, we envision extending this work in another direction. Since the prominence and versatility of Large Language Models (LLMs) [99, 104] have reached unprece-dented heights, an additional future work could be to embed LLMs (i.e., GPT-4 and Llama 2) into the human-in-the-loop interaction stage to assist human experts in the selection of the routine segments to be filtered out.

The limitation of the envisioned segmentation approach is that it assumes every routine captured in a UI log has clear endpoints (i.e., the end-delimiters of a routine). If such points do not exist, there is no strictly defined way of performing the segmentation task, since we may discover incorrect routine segments, thus affecting the quality of the discovered routines.

9.2 Overview on the Automated Generation of SW Robots

Although RPA is currently used for automating routines and high-volume tasks requiring manual intervention of expert users, SmartRPA aims to automatically develop SW robots directly from the users' observed behaviour, thus tackling **C2**. SmartRPA offers an innovative contribution to RPA technology to mitigate some of its core downsides. We leveraged a design science research method [52] to build the SmartRPA approach, which is able to interpret the UI logs keeping track of many routine executions, and to automatically synthesize SW robots that emulate the most suitable routine variant for any specific intermediate user input that is required during the routine execution.

Notably, using SmartRPA, all the routine executions recorded by the tool can be automated, a high-level flowchart diagram is presented to expert users for potential diagnosis operations, and the executable RPA scripts to drive the working of a SW robot are generated by solely interpreting the routine executions stored in the routine-based log, selecting step-by-step the most suitable routine variant.

From a technical perspective, the script generation algorithm takes into account only the platform where the SW robot is going to be run, regardless of the operating system used to capture the log. For example, if the selected routine variant was recorded on MacOS, but the tool is being executed on Windows, the RPA script will be generated taking into account this aspect, e.g., by converting the information about the system paths. This guarantees cross-platform compatibility across UI logs recorded on different platforms, as suggested by the guidelines' principles of RPM. Last but not least, SmartRPA creates executable RPA scripts also for UiPath, one of the major vendors in the RPA market. These scripts can then be executed via the interface of UiPath. In addition, the tool allows us to personalize some input fields of the selected routine variant before executing the related RPA scripts (either on Windows/MacOS systems or within UiPath Studio), thus supporting those steps that require intermediate manual user inputs. As a consequence, this makes the working of SW robots flexible and adaptable to several real-world situations. To sum up, we consider SmartRPA as an important first step towards the intelligent fully automated generation of SW robots.

The main weakness of the approach is correlated with the quality of information recorded in real-world UI logs. Since a UI log is fine-grained, routines executed with many different strategies may potentially affect the *robustness* of our approach to the detection of variation points. For this reason, as future work, we are going to perform a robust evaluation of the algorithm on further real-world case studies including heterogeneous UI logs obtained from different application domains.

Apart from the ability to automatically generate the SW robots' behaviour, thanks to its Action Logger, SmartRPA aims also at improving the *auditability* of RPA tools since all routines executed by human users on a UI are previously recorded in dedicated event logs, making them auditable to external users. The logs produced by the state-of-the-art RPA tools have usually poor quality (actions may be missing or not recorded properly), since they are mainly used for debugging purposes [10]. Conversely, SmartRPA aims at logs at the highest possible quality level thanks to its detailed recording phase performed during the training sessions.

Moreover, we also envision that the proposed book will provide long-term benefits to the company's workforce. With entry-level and repetitive jobs mostly performed by SW robots, the majority of resources can be reassigned to more rewarding activities and, consequently, job satisfaction will increase. Furthermore, *scalability* will be improved as well. Human capacity is difficult to scale in situations where demand fluctuates, leading to inefficiencies such as backlogs or overcapacity. In contrast, SW robots generated by SmartRPA operate at whatever speed is demanded by the work volume.

9.3 Final Considerations

The research underlying this book can be considered in the frame of the European initiatives for *Industry 5.0*, toward the development of a more sustainable, resilient, and human-centric industry leveraging automated intelligent techniques and Industry 4.0 (see `https://tinyurl.com/4vhs282f`). Indeed, this work will constitute an additional contribution to the comprehensive approach that exploits intelligent technologies to empower the abilities of human resources, a relevant objective highlighted in recent calls of the *Horizon EU Framework* (see `https://tinyurl.com/2p8fhs9j`).

References

1. L. Abb, C. Bormann, H. van der Aa, and J. Rehse. Trace Clustering for User Behavior Mining. In *30th European Conference on Information Systems (ECIS 2022)*, 2022.
2. A. Adriansyah, N. Sidorova, and B. F. van Dongen. Cost-Based Fitness in Conformance Checking. In *Int. Conf. on Application of Concurrency to System Design*, pages 57–66. IEEE, 2011.
3. S. Agostinelli, F. Leotta, and A. Marrella. Interactive Segmentation of User Interface Logs. In *19th Int. Conf. on Service-Oriented Computing (ICSOC 2021)*, volume 13121, pages 65–80, 2021.
4. S. Agostinelli, M. Lupia, A. Marrella, and M. Mecella. Automated Generation of Executable RPA Scripts from User Interface Logs. In *Blockchain and Robotic Process Automation Forum (BPM 2020)*, volume 393, pages 116–131. Springer, 2020.
5. S. Agostinelli, M. Lupia, A. Marrella, and M. Mecella. SmartRPA: A Tool to Reactively Synthesize Software Robots from User Interface Logs. In *33rd Int. Conf. on Advanced Information Systems Engineering (CAiSE Forum 2021)*, volume 424, pages 137–145. Springer, 2021.
6. S. Agostinelli, M. Lupia, A. Marrella, and M. Mecella. Reactive Synthesis of Software Robots in RPA from User Interface Logs. *Computers in Industry*, 142:103721, 2022.
7. S. Agostinelli, F. M. Maggi, A. Marrella, and F. Milani. A User Evaluation of Process Discovery Algorithms in a Software Engineering Company. In *2019 IEEE 23rd International Enterprise Distributed Object Computing Conference (EDOC)*, pages 142–150, 2019.
8. S. Agostinelli, A. Marrella, L. Abb, and J. Rehse. Mastering Robotic Process Automation with Process Mining. In *20th Int. Conf. Business Process Management (BPM 2022)*, volume 13420, pages 47–53. Springer, 2022.
9. S. Agostinelli, A. Marrella, and M. Mecella. Research Challenges for Intelligent Robotic Process Automation. In *Business Process Management Workshops (BPM 2019)*, pages 12–18, 2019.
10. S. Agostinelli, A. Marrella, and M. Mecella. Towards Intelligent Robotic Process Automation for BPMers, 2020.
11. S. Agostinelli, A. Marrella, and M. Mecella. Automated Segmentation of User Interface Logs. In *Robotic Process Automation: Management, Technology, Applications*, pages 201–222. De Gruyter, 2021.
12. S. Aguirre and A. Rodriguez. Automation of a Business Process Using Robotic Process Automation (RPA): A Case Study. In *Applied Computer Sciences in Engineering*, pages 65–71. Springer, 2017.
13. AI-Multiple. Top 53 RPA Tools / Vendors & Their Features, 2024.
14. S. Anagnoste. Setting up a Robotic Process Automation Center of Excellence. *Management Dynamics in the Knowledge Economy*, 6(2):307–332, 2018.
15. A. Augusto, R. Conforti, M. Dumas, M. La Rosa, F. M. Maggi, A. Marrella, M. Mecella, and A. Soo. Automated Discovery of Process Models from Event Logs: Review and Benchmark. *IEEE Trans. Knowl. Data Eng.*, 31(4):686–705, 2019.
16. A. Ayub and A. R. Wagner. Teach Me What You Want to Play: Learning Variants of Connect Four through Human-Robot Interaction, 2020.
17. T. Baier, A. Rogge-Solti, J. Mendling, and M. Weske. Matching of Events and Activities: An Approach Based on Behavioral Constraint Satisfaction. In *ACM Symp. on Applied Computing*, pages 1225–1230, 2015.
18. D. Bayomie, C. D. Ciccio, M. La Rosa, and J. Mendling. A Probabilistic Approach to Event-Case Correlation for Process Mining. In *38th Int. Conf. on Conceptual Modeling (ER'19)*, volume 11788, pages 136–152, 2019.
19. Y. Benjamini and Y. Hochberg. Controlling the False Discovery Rate: A Practical and Powerful Approach to Multiple Testing. *Journal of the Royal statistical society: series B (Methodological)*, 57(1):289–300, 1995.

20. A. Berti, S. J. van Zelst, and W. van der Aalst. Process Mining for Python (PM4Py): Bridging the Gap Between Process- and Data Science, 2019.
21. H. H. Beyel and W. M. P. van der Aalst. Creating Translucent Event Logs to Improve Process Discovery. In *Int. Conf. on Process Mining (ICPM'22)*, pages 435–447. Springer, 2022.
22. A. Bosco, A. Augusto, M. Dumas, M. La Rosa, and G. Fortino. Discovering Automatable Routines From User Interaction Logs. In *Int. Conf. on Business Process Management (BPM'19), Forum track, Vienna, Austria*, pages 144–162. Springer, 2019.
23. J. Brooke. SUS: A Retrospective. *Journal of Usability Studies*, 8(2):29–40, 2013.
24. J. C. Campos, M. Sousa, M. C. B. Alves, and M. D. Harrison. Formal Verification of a Space System's User Interface with the IVY Workbench. *IEEE SMC*, 46(2):303–316, 2016.
25. T. Chakraborti, V. Isahagian, R. Khalaf, Y. Khazaeni, V. Muthusamy, Y. Rizk, and M. Unuvar. From Robotic Process Automation to Intelligent Process Automation: Emerging Trends. In *Business Process Management: Blockchain and Robotic Process Automation Forum - BPM '20*, pages 215–228. Springer, 2020.
26. I. M. Chakravarti, R. G. Laha, and J. Roy. Handbook of Methods of Applied Statistics. *Wiley Series in Probability and Mathematical Statistics (USA)*, 1967.
27. D. J. Cook, N. C. Krishnan, and P. Rashidi. Activity Discovery and Activity Recognition: A New Partnership. *IEEE Transactions on Cybernetics*, 43(3):820–828, 2013.
28. M. de Leoni, M. Dumas, and L. García-Bañuelos. Discovering Branching Conditions from Business Process Execution Logs. In *16th International Conference on Fundamental Approaches to Software Engineering (FASE'13)*, pages 114–129, 2013.
29. M. de Leoni, G. Lanciano, and A. Marrella. Aligning Partially-Ordered Process-Execution Traces and Models Using Automated Planning. In *28th Int. Conf. on Automated Planning and Scheduling (ICAPS 2018)*, pages 321–329, 2018.
30. M. de Leoni and A. Marrella. Aligning Real Process Executions and Prescriptive Process Models through Automated Planning. *Expert System with Application*, 82:162–183, 2017.
31. C. Di Ciccio, A. Marrella, and A. Russo. Knowledge-Intensive Processes: Characteristics, Requirements and Analysis of Contemporary Approaches. *Journal on Data Semantics*, 4(1):29–57, Mar 2015.
32. M. Dignum. *A Model for Organizational Interaction: Based on Agents, Founded in Logic.* SIKS, 2004.
33. A. Dix, J. Finlay, G. Abowd, and R. Beale. Human-Computer Interaction. *Pearson*, 2004.
34. Dix, Alan. Statistics for HCI: Making Sense of Quantitative Data. *Synthesis Lectures on Human-Centered Informatics*, 13(2):1–181, 2020.
35. M. Dumas, F. Fournier, L. Limonad, A. Marrella, M. Montali, J.-R. Rehse, R. Accorsi, D. Calvanese, G. De Giacomo, D. Fahland, et al. AI-Augmented Business Process Management Systems: A Research Manifesto. *ACM Transactions on Management Information Systems*, 14(1):1–19, 2023.
36. M. Dumas, M. L. Rosa, J. Mendling, and H. A. Reijers. *Fundamentals of Business Process Management*. Springer, 2013.
37. B. Fazzinga, S. Flesca, F. Furfaro, E. Masciari, and L. Pontieri. Efficiently Interpreting Traces of Low Level Events in Business Process Logs. *Information Systems*, 73:1–24, 2018.
38. M. S. Feary. A Toolset for Supporting Iterative Human Automation: Interaction in Design. *NASA Ames Research Center*, 2010.
39. D. R. Ferreira, F. Szimanski, and C. G. Ralha. Improving Process Models by Mining Mappings of Low-Level Events to High-Level Activities. *Intelligent Information Systems*, 43(2):379–407, 2014.
40. J. Gao, S. J. van Zelst, X. Lu, and W. M. P. van der Aalst. Automated Robotic Process Automation: A Self-Learning Approach. In *On the Move to Meaningful Internet Systems: OTM 2019 Conf.*, pages 95–112. Springer, 2019.
41. J. Geyer-Klingeberg, J. Nakladal, F. Baldauf, and F. Veit. Process Mining and Robotic Process Automation: A Perfect Match. In *16th Int. Conf. on Business Process Management (BPM'18), Dissertation/Demos/Industry track*, 2018.

42. A. F. Ghahfarokhi, G. Park, A. Berti, and W. M. van der Aalst. OCEL: A Standard for Object-Centric Event Logs. In *European Conference on Advances in Databases and Information Systems*, pages 169–175. Springer, 2021.

43. C. W. Günther and E. Verbeek. XES Standard Definition. *Fluxicon Process Laboratories*, 2014.

44. X. Han, L. Hu, Y. Dang, S. Agarwal, L. Mei, S. Li, and X. Zhou. Automatic Business Process Structure Discovery using Ordered Neurons LSTM: A Preliminary Study. *arXiv CoRR abs/2001.01243*, 2020.

45. D. Harel. Statecharts: A Visual Formalism for Complex Systems. *Science of computer programming*, 8(3):231–274, 1987.

46. L.-V. Herm, C. Janiesch, A. Helm, F. Imgrund, K. Fuchs, A. Hofmann, and A. Winkelmann. A Consolidated Framework for Implementing Robotic Process Automation Projects. In *18th Int. Conf. on Business Process Management, BPM'20*, pages 471–488. Springer, 2020.

47. IEEE Digital Library. Standard for eXtensible Event Stream (XES) for Achieving Interoperability in Event Logs and Event Streams. *IEEE Std 1849-2016*, 2016.

48. F. Imgrund, M. Fischer, C. Janiesch, and A. Winkelmann. Conceptualizing a Framework to Manage the Short Head and Long Tail of Business Processes. In *16th Int. Conf. on Business Process Management, BPM'18*, pages 392–408. Springer, 2018.

49. N. Ito, Y. Suzuki, and A. Aizawa. From Natural Language Instructions to Complex Processes: Issues in Chaining Trigger Action Rules. *arXiv CoRR abs/2001.02462*, 2020.

50. P. Jenkins, H. Wei, J. S. Jenkins, and Z. Li. A Probabilistic Simulator of Spatial Demand for Product Allocation, 2020.

51. A. Jimenez-Ramirez, H. A. Reijers, I. Barba, and C. Del Valle. A Method to Improve the Early Stages of the Robotic Process Automation Lifecycle. In *31st Int. Conf. on Advanced Information Systems Engineering (CAiSE'19)*, pages 446–461, 2019.

52. P. Johannesson and E. Perjons. *An Introduction to Design Science*. Springer, 2014.

53. B. E. John and D. E. Kieras. The GOMS Family of User Interface Analysis Techniques: Comparison and Contrast. *ACM TOCHI*, 3(4), 1996.

54. M. Kirchmer. Robotic Process Automation-Pragmatic Solution or Dangerous Illusion. *BTOES Insights, June'17*, 2017.

55. J. Kokina and S. Blanchette. Early Evidence of Digital Labor in Accounting: Innovation with Robotic Process Automation. *International Journal of Accounting Information Systems*, 35, 2019.

56. A. Kumar, J. Salo, and H. Li. Stages of User Engagement on Social Commerce Platforms: Analysis with the Navigational Clickstream Data. *Int. J. El. C.*, 23(2), 2019.

57. M. Lacity, L. P. Willcocks, and A. Craig. *RPA at Telefonica O2*. The London School of Economics and Political Science, 2015.

58. C. Langmann, D. Turi, et al. *Robotic Process Automation (RPA)-Digitalisierung und Automatisierung von Prozessen*. Springer Books, 2020.

59. V. Le and S. Gulwani. FlashExtract: A Framework for Data Extraction by Examples. In *ACM SIGPLAN PLDI '14*, pages 542–553, 2014.

60. V. Leno, A. Augusto, M. Dumas, M. La Rosa, F. M. Maggi, and A. Polyvyanyy. Identifying Candidate Routines for Robotic Process Automation from Unsegmented UI Logs. In *2nd Int. Conf. on Process Mining*, pages 153–160, 2020.

61. V. Leno, S. Deviatykh, A. Polyvyanyy, M. L. Rosa, M. Dumas, and F. M. Maggi. Robidium: Automated Synthesis of Robotic Process Automation Scripts from UI Logs. In *Demonstration & Resources Track at 18th International Conference on Business Process Management (BPM 2020)*, pages 102–106, 2020.

62. V. Leno, M. Dumas, F. Maggi, and M. La Rosa. Multi-Perspective Process Model Discovery for Robotic Process Automation. In *CEUR Workshop Proceedings*, volume 2114, pages 37–45, 2018.

63. V. Leno, A. Polyvyanyy, M. Dumas, M. L. Rosa, and F. M. Maggi. Robotic Process Mining: Vision and Challenges. *Bus. Inf. Syst. Eng.*, 63(3):301–314, 2021.

64. V. Leno, A. Polyvyanyy, M. L. Rosa, M. Dumas, and F. M. Maggi. Action Logger: Enabling Process Mining for Robotic Process Automation. In *Proceedings of the Dissertation Award, Doctoral Consortium, and Demonstration Track at 17th Int. Conf. on Business Process Management, (BPM'19)*, pages 124–128, 2019.

65. H. Leopold, H. van der Aa, and H. A. Reijers. Identifying Candidate Tasks for Robotic Process Automation in Textual Process Descriptions. In *Int. Conf. on Bus. Proc. Mod., Dev. and Supp. (BPMDS'18)*, pages 67–81. Springer, 2018.

66. F. Leotta, M. Mecella, D. Sora, and T. Catarci. Surveying Human Habit Modeling and Mining Techniques in Smart Spaces. *Future Internet*, 11(1):23, 2019.

67. C. Linn, P. Zimmermann, and D. Werth. Desktop Activity Mining - A New Level of Detail in Mining Business Processes. In *Workshops der INFORMATIK 2018 - Architekturen, Prozesse, Sicherheit und Nachhaltigkeit, 26.-27*, pages 245–258, 2018.

68. X. Liu. Unraveling and Learning Workflow Models from Interleaved Event Logs. In *2014 IEEE Int. Conf. on Web Services*, pages 193–200, 2014.

69. S. Lohr. The Beginning of a Wave: A.I. Tiptoes Into the Workplace. https://www.nytimes.com/2018/08/05/technology/workplace-ai.html/, 2018.

70. F. Mannhardt, M. de Leoni, H. A. Reijers, W. M. van der Aalst, and P. J. Toussaint. Guided Process Discovery – A Pattern-Based Approach. *Inf. Syst.*, 76:1–18, 2018.

71. A. Marrella. What Automated Planning Can Do for Business Process Management. In *BPM 2017 International Workshops, AI4BBPM'17*, 2017.

72. A. Marrella. Automated Planning for Business Process Management. *J. Data Semantic*, 8(2):79–98, 2019.

73. A. Marrella and T. Catarci. Measuring the Learnability of Interactive Systems Using a Petri Net Based Approach. In *2018 Designing Interactive Systems Conf.*, DIS '18, pages 1309–1319. ACM, 2018.

74. A. Marrella, M. Mecella, and S. Sardiña. Supporting Adaptiveness of Cyber-Physical Processes through Action-Based Formalisms. *AI Commun.*, 31(1):47–74, 2018.

75. A. Martínez-Rojas, A. Jiménez-Ramírez, J. G. Enríquez, and H. A. Reijers. Analyzing Variable Human Actions for Robotic Process Automation. In *20th Int. Conf. on Business Process Management (BPM'22)*, volume 13420, pages 75–90. Springer, 2022.

76. A. Martínez-Rojas, A. Jiménez-Ramírez, J. G. Enríquez, and H. A. Reijers. A screenshot-based task mining framework for disclosing the drivers behind variable human actions. *Information Systems*, 121:102340, 12 2023.

77. L. Măruşter, A. T. Weijters, W. M. Van Der Aalst, and A. Van Den Bosch. A Rule-Based Approach for Process Discovery: Dealing with Noise and Imbalance in Process Logs. *Data Mining and Knowledge Discovery*, 13(1):67–87, 2006.

78. A. Meironke and S. Kuehnel. How to Measure RPA's Benefits? A Review on Metrics, Indicators, and Evaluation Methods of RPA Benefit Assessment. In *17th Int. Tagung Wirtschaftsinformatik (WI'22)*. AISeL, 2022.

79. A. Miltner, S. Gulwani, V. Le, A. Leung, A. Radhakrishna, G. Soares, A. Tiwari, and A. Udupa. On the Fly Synthesis of Edit Suggestions. *In: ACM Program. Lang.*, 3(OOPSLA):143:1–143:29, 2019.

80. M. Montali, M. Pesic, W. M. P. van der Aalst, F. Chesani, P. Mello, and S. Storari. Declarative Specification and Verification of Service Choreographies. *ACM Transactions on the Web*, 4(1), 2010.

81. G. Mori, F. Paternò, and C. Santoro. CTTE: Support for Developing and Analyzing Task Models for Interactive System Design. *IEEE Trans. Sof. Eng.*, 28(8), 2002.

82. P. A. Palanque and R. Bastide. Petri Net Based Design of User-Driven Interfaces Using the Interactive Cooperative Objects Formalism. In *Interactive Systems: Design, Specification, and Verification*. Springer, 1995.

83. F. Paternò. *Model-Based Design and Evaluation of Interactive Applications*. Springer-Verlag, 1st edition, 1999.

84. E. Penttinen, H. Kasslin, and A. Asatiani. How to Choose between Robotic Process Automation and Back-end System Automation? In *European Conference on Information Systems (ECIS 2018)*, 2018.

85. A. Pnueli. The Temporal Logic of Programs. In *F. of Comp. Sc.*, 1977.
86. R. Ravn, P. Halberg, J. Gustafsson, and J. Groes. Get Ready for Robots: Why Planning Makes the Difference Between Success and Disappointment. https://eyfinancialservicesthoughtgallery.ie/wp-content/uploads/2016/11/ey-get-ready-for-robots.pdf, 2016. Accessed: 19-07-2021.
87. A. Rebmann and H. van der Aa. Unsupervised Task Recognition from User Interaction Streams. In *35th Int. Conf. on Advanced Information Systems Engineering (CAiSE 2023)*, volume 13901, pages 141–157. Springer, 2023.
88. M. Rosemann and J. v. Brocke. *Handbook on Business Process Management 1: Introduction, Methods, and Information Systems*. Springer, 2 edition, 2015.
89. M. Rovani, F. M. Maggi, M. de Leoni, and W. M. van der Aalst. Declarative Process Mining in Healthcare. *Expert Systems with Applications*, 42(23), 2015.
90. J. Sauro and J. R. Lewis. *Quantifying the User Experience: Practical Statistics for User Research*. Morgan Kaufmann, 2016.
91. M. Schmitz, C. Dietze, and C. Czarnecki. Enabling Digital Transformation through Robotic Process Automation at Deutsche Telekom. In *Digitalization Cases*, pages 15–33. Springer, 2019.
92. M. Smeets, R. Erhard, T. Kaußler, et al. *Robotic Process Automation (RPA) in der Finanzwirtschaft*. Springer Books, 2019.
93. J. Srivastava, R. Cooley, M. Deshpande, and P. Tan. Web Usage Mining: Discovery and Applications of Usage Patterns from Web Data. *SIGKDD Exp.*, 1(2), 2000.
94. A. G. Sutcliffe and I. Wang. Integrating Human Computer Interaction with Jackson System Development. *The Computer journal*, 34(2), 1991.
95. O. Sy, R. Bastide, P. Palanque, D. Le, and D. Navarre. PetShop: A CASE Tool for the Petri Net Based Specification and Prototyping of CORBA Systems. In *Petri nets*, volume 2000, page 78. Citeseer, 2000.
96. C. Tornbohm and R. Dunie. Market Guide for Robotic Process Automation Software. *Gartner*, 2017.
97. Y. Urabe, S. Yagi, K. Tsuchikawa, and H. Oishi. Task Clustering Method using User Interaction Logs to Plan RPA Introduction. In *19th Int. Conf. on Business Process Management (BPM 2021)*, pages 273–288. Springer, 2021.
98. J. van den Bos, M. J. Plasmeijer, and P. H. Hartel. Input-Output Tools: A Language Facility for Interactive and Real-Time Systems. *IEEE Trans. Software Eng.*, 9(3):247–259, 1983.
99. H. van der Aa, J. Carmona, H. Leopold, J. Mendling, and L. Padró. Challenges and Opportunities of Applying Natural Language Processing in Business Process Management. In *27th International Conference on Computational Linguistics, COLING 2018*, pages 2791–2801. Association for Computational Linguistics, 2018.
100. W. van der Aalst, M. Pesic, and H. Schonenberg. Declarative Workflows: Balancing Between Flexibility and Support. *Computer Science - R&D*, pages 99–113, 2009.
101. W. M. van der Aalst. Process Mining: A 360 Degree Overview. In *Process Mining Handbook*, pages 3–34. Springer, 2022.
102. W. M. P. van der Aalst. *Process Mining - Data Science in Action, Second Edition*. Springer, 2016.
103. W. M. P. van der Aalst, M. Bichler, and A. Heinzl. Robotic Process Automation. *Bus. Inf. Syst. Eng.*, 60(4):269–272, 2018.
104. M. Vidgof, S. Bachhofner, and J. Mendling. Large Language Models for Business Process Management: Opportunities and Challenges. In *Business Process Management Forum (BPM 2023)*, volume 490, pages 107–123. Springer, 2023.
105. A. I. Wasserman. Extending State Transition Diagrams for the Specification of Human-Computer Interaction. *IEEE Trans. Software Eng.*, 11(8):699–713, 1985.
106. M. Weske. *Business Process Management: Concepts, Languages, Architectures*. Springer, 2 edition, 2019.